KW-050-328

PROPERTY OF
SURFACE PHYSICS GROUP,
DEPARTMENT OF PHYSICS,
UNIVERSITY OF YORK.

Springer Tracts in Modern Physics 91

Editor: G. Höhler
Associate Editor: E. A. Niekisch

Editorial Board: S. Flügge H. Haken J. Hamilton
H. Lehmann W. Paul

Springer Tracts in Modern Physics

68* **Solid-State Physics**   With contributions by D. Bäuerle, J. Behringer, D. Schmid

69* **Astrophysics**   With contributions by G. Börner, J. Stewart, M. Walker

70* **Quantum Statistical Theories of Spontaneous Emission and their Relation to Other Approaches**   By G. S. Agarwal

71   **Nuclear Physics**   With contributions by J. S. Levinger, P. Singer, H. Überall

72   **Van der Waals Attraction:**   Theory of Van der Waals Attraction By D. Langbein

73   **Excitons at High Density**   Edited by H. Haken, S. Nikitine. With contributions by V. S. Bagaev, J. Biellmann, A. Bivas, J. Goll, M. Grosmann, J. B. Grun, H. Haken, E. Hanamura, R. Levy, H. Mahr, S. Nikitine, B. V. Novikov, E. I. Rashba, T. M. Rice, A. A. Rogachev, A. Schenzle, K. L. Shaklee

74   **Solid-State Physics**   With contributions by G. Bauer, G. Borstel, H. J. Falge, A. Otto

75   **Light Scattering by Phonon-Polaritons**   By R. Claus, L. Merten, J. Brandmüller

76   **Irreversible Properties of Type II Superconductors**   By H. Ullmaier

77   **Surface Physics**   With contributions by K. Müller, P. Wißmann

78   **Solid-State Physics**   With contributions by R. Dornhaus, G. Nimtz, W. Richter

79   **Elementary Particle Physics**   With contributions by E. Paul, H. Rollnick, P. Stichel

80* **Neutron Physics**   With contributions by L. Koester, A. Steyerl

81   **Point Defects in Metals I:**   Introduction to the Theory 2nd Printing By G. Leibfried, N. Breuer

82   **Electronic Structure of Noble Metals, and Polariton-Mediated Light Scattering** With contributions by B. Bendow, B. Lengeler

83   **Electroproduction at Low Energy and Hadron Form Factors** By E. Amaldi, S. P. Fubini, G. Furlan

84   **Collective Ion Acceleration**   With contributions by C. L. Olson, U. Schumacher

85   **Solid Surface Physics**   With contributions by J. Hölzl, F. K. Schulte, H. Wagner

86   **Electron-Positron Interactions**   By B. H. Wiik, G. Wolf

87   **Point Defects in Metals II:**   Dynamical Properties and Diffusion Controlled Reactions With contributions by P. H. Dederichs, K. Schroeder, R. Zeller

88   **Excitation of Plasmons and Interband Transitions by Electrons**   By H. Raether

89   Giant Resonance Phenomena in **Intermediate-Energy Nuclear Reactions** By F. Cannata, H. Überall

90* **Jets of Hadrons**   By W. Hofmann

91   **Structural Studies of Surfaces** With contributions by K. Heinz, K. Müller, T. Engel, and K. H. Rieder

92   **Single-Particle Rotations in Molecular Crystals**   By W. Press

93   **Coherent Inelastic Neutron Scattering in Lattice Dynamics**   By B. Dorner

94   **Exciton Dynamics in Molecular Crystals and Aggregates**   With contributions by V. M. Kenkre and P. Reineker

* denotes a volume which contains a Classified Index starting from Volume 36.

Structural Studies
of Surfaces

Contributions by
K. Heinz   K. Müller   T. Engel   K.-H. Rieder

With 120 Figures

Springer-Verlag
Berlin Heidelberg New York 1982

Professor Dr. Klaus Heinz
Professor Dr. Klaus Müller
Institut für Angewandte Physik, Universität Erlangen,
Erwin-Rommel-Straße 1, D-8520 Erlangen, Fed. Rep. of Germany

Dr. Thomas Engel
Department of Chemistry, University of Washington, Seattle, WA 98195, USA
Dr. Karl-Heinz Rieder
IBM Zurich Research Laboratory, CH-8803 Rüschlikon, Switzerland

---

Manuscripts for publication should be addressed to:

Gerhard Höhler
Institut für Theoretische Kernphysik der Universität Karlsruhe
Postfach 6380, D-7500 Karlsruhe 1, Fed. Rep. of Germany

*Proofs and all correspondence concerning papers in the process of publication
should be addressed to:*

Ernst A. Niekisch
Haubourdinstrasse 6, D-5170 Jülich, Fed. Rep. of Germany

ISBN 3-540-10964-1 Springer-Verlag Berlin Heidelberg New York
ISBN 0-387-10964-1 Springer-Verlag New York Heidelberg Berlin

Library of Congress Cataloging in Publication Data. Main entry under title: Structural studies of surfaces.
(Springer tracts in modern physics; 91). Includes bibliographical references and index. Contents: LEED intensities:
 experimental progress and new possibilities of surface structure determination / by K. Heinz and K. Müller –
Structural studies of surfaces with atomic and molecular beam diffraction / T. Engel and K.-H. Rieder. 1. Surfaces
(Physics) – Addresses, essays, lectures. 2. Diffraction – Addresses, essays, lectures. I. Heinz, Klaus. II. Series QCI.S797.
vol. 91 [QC173.4.S94]   539s[530.4]   81-9327AACR2

This work is subject to copyright. All rights are reserved, whether the whole or part of the material is
concerned, specifically those of translation, reprinting, reuse of illustrations, broadcasting, reproduction by
photocopying machine or similar means, and storage in data banks. Under § 54 of the German Copyright
Law where copies are made for other than private use, a fee is payable to „Verwertungsgesellschaft Wort", Munich.

© by Springer-Verlag Berlin Heidelberg 1982
Printed in Germany

The use of registered names, trademarks, etc. in this publication does not imply, even in the absence
of a specific statement, that such names are exempt from the relevant protective laws and regulations
and therefore free for general use.

Offset printing and bookbinding: Brühlsche Universitätsdruckerei, Giessen
2153/3130 − 5 4 3 2 1 0

Contents

LEED Intensities – Experimental Progress and New Possibilities of Surface Structure Determination

By *K. Heinz* and *K. Müller*. With 29 Figures

1. Introduction 1

2. Structural Determination by Comparison of Experimental and Calculated
   Intensities 3
   2.1 Comparison of Theoretical and Experimental Spectra 4
   2.2 Comparison of Experimental Spectra from Different Measurements 7
   2.3 Sources of Experimental Error 9
   2.4 Classical Methods for Intensity Measurements 13

3. New Experimental Methods for Intensity Data Collection 18
   3.1 Photographic Methods 18
   3.2 TV Computer Methods 20
        3.2.1 Data Acquisition Rate Lower than TV Rate 21
        3.2.2 Data Acquisition Rate Equal to TV Rate 27

4. Examples for Reliable Intensity Data Obtained by the New Methods 33
   4.1 Influence of Sample Misalignment 33
   4.2 Influence of Background Subtraction 36
   4.3 Influence of Residual Gas Adsorption 38
   4.4 Influence of Adsorbate Decomposition and Desorption 38

5. New Possibilities Using Modern Intensity Measurement Methods 39
   5.1 Integral Intensities of Rapidly Varying Surface Systems 39
   5.2 Spot Profiles of Rapidly Varying Surface Systems 42
   5.3 Extension of Intensity Measurements to Varying Temperature 43
   5.4 Extension of Intensity Measurements to the Medium-Energy Range 45

6. Summary and Outlook 48
References 49

Structural Studies of Surfaces with Atomic and Molecular Beam Diffraction

By *T. Engel* and *K.-H. Rieder*. With 91 Figures

1. Introduction 55

2. The Particle-Surface Interaction Potential 57
   2.1 Physical Basis 57
   2.2 Short Survey of Theoretical Efforts 58
   2.3 Determination of the Surface Potential from Bound-State Energy Data   65

3. Quantum Theory of Particle Diffraction 71
   3.1 The Corrugated Hard-Wall Model 71
   3.2 Diffraction Condition and Ewald Construction 72
   3.3 Calculation of Diffraction Intensities — General Method 75
   3.4 Calculation of Intensities — Rayleigh Hypothesis 78
       3.4.1 The GR Method 79
       3.4.2 The Eikonal Approximation 79
   3.5 Calculation of Intensities — Iterative Series 83
   3.6 A Few Illustrative Examples 86
   3.7 The Inversion Problem 90
   3.8 Effects Due to the Softness of the Repulsive Potential 92

4. Inelastic Scattering of Atoms from Surfaces 93
   4.1 The Dependence of the Scattering on the Time Scale of the Interaction   93
   4.2 The Debye-Waller Factor in the Time-Dependent Interaction Regime   95
   4.3 The Size Effect in the Debye-Waller Factor for Atom Scattering   97
   4.4 Experimental Investigations of the Debye-Waller Factor for Atom-
       Surface Scattering 98

5. Influence of the Attractive Part of the Potential on Diffraction Intensities 103
   5.1 Modifications for the Calculation of Diffraction Intensities 103
   5.2 Bound Surface States and Resonant Transitions 105
   5.3 Theory of Atom Scattering from a Corrugated Hard Wall with an
       Attractive Well 111
   5.4 Inelastic Effects in Resonant Scattering 114

6. Experimental Aspects of Gas-Surface Scattering 118
   6.1 Requirements on an Apparatus to Perform Gas-Surface Scattering
       Experiments 118

6.2   Beam Sources 119
      6.2.1 Effusive Beam Sources 119
      6.2.2 Nozzle-Beam Sources 119
6.3   Beam Energy Variation for Effusive and Nozzle-Beam Sources 123
6.4   The Design of Nozzle-Beam Systems 124
6.5   Molecular-Beam Detectors 126
6.6   Detector Rotation 128
6.7   Sample Manipulators 129
6.8   Beam-Modulation Devices 131
6.9   Experimental Systems for Diffractive Scattering from Surfaces 133
6.10 The Influence of the Transfer Width of the Apparatus and of Surface
      Perfection on Measured Intensities 137

7. Structural Investigations on Surfaces of Ionic Crystals 140
   7.1 Diffraction Studies on LiF(100) 140
   7.2 Diffraction Studies on NiO(100) 146
   7.3 Diffraction from Other Ionic Materials 147

8. Structural Investigations on Semiconductor Surfaces 148
   8.1 Helium-Diffraction Studies on Si(111) and Si(100) 148
   8.2 Helium Diffraction from GaAs(110) 152
   8.3 Diffraction from Graphite 154
   8.4 Diffraction from Layer Compounds 155
   8.5 Helium-Diffraction Studies from Other Surfaces 156

9. Structural Investigations on Metal Surfaces 157
   9.1 Introduction 157
   9.2 Helium and Hydrogen Diffraction from Close-Packed Metal Surfaces 157
   9.3 Helium Diffraction from fcc(110) and bcc(112) Planes 158
   9.4 Helium Diffraction from Stepped Metal Surfaces 160

10. Structural Studies on Adsorbate-Covered Surfaces 162
    10.1 Introduction 162
    10.2 Hydrogen Adsorption on Ni(110) 163
    10.3 Oxygen Adsorption on Ni(110) 171
    10.4 Oxygen Adsorption on Cu(110) 173

References 174

LEED Intensities – Experimental Progress and New Possibilities of Surface Structure Determination

K. Heinz and K. Müller

1. Introduction

The structure of a crystal surface is one of its important properties. Other quantities, such as the surface density of states, the electronic work function, and chemical bonding or chemical reactions in the presence of adsorbates, are correlated to structure parameters. It is therefore one of the basic aims of surface science to investigate the surface structure, i.e., the geometric arrangement of atoms in the first few layers of a crystal.

Among the techniques sensitive to surface structure is diffraction of low-energy particles with De Broglie wavelenghts of the order of the lattice constant resulting from ionic scattering (e.g., /1.1-3/) and atomic or molecular scattering (e.g., /1.4,5/). Methods more sensitive to local order, such as angular-resolved photoemission spectroscopy (e.g., /1.6,7/) or angular resolved Auger-electron spectroscopy (e.g., /1.8-13/), are also in progress. Low-energy electron diffraction (LEED), however, is the oldest and most commonly used method for surface structure determination (e.g., /1.14-30/). A more or less coherent beam of electrons in the energy range of about 20-600 eV, i.e., with wavelengths between 3 Å and 0.5 Å, impinges on the crystalline surface. The elastically backscattered electrons, separated from an inelastic background by retarding field grids, display the diffraction pattern of the surface on a luminescent screen. Both electron-beam generation by an electron gun and detection of the diffraction pattern can be easily performed by commercially available UHV equipment.

In spite of the simplicity of the experimental arrangement, only since about 1974 have the number of surface structure determinations by LEED been notably increasing as shown in Fig.1.1. The reason for this comparatively late start is the complexity of low-energy electron scattering. The geometry of the diffraction pattern is determined by the translational symmetry and the periodicity of the surface layers only, and so shape and size of the surface unit cell can be deduced by simple arguments. However, essential details of the unit cell, i.e., the geometric arrange-

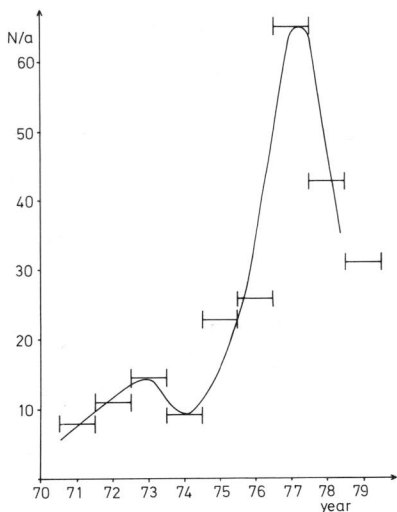

Fig.1.1. Development of the number of publications appearing per year on surface-
structure determination by LEED during the
last decade as taken from the Surface and
Vacuum Physics Index

ment of its atoms, can be determined only by independent model calculations of the
intensities of diffracted beams and subsequent comparison with the experiment. This
holds also for the mutual orientation between adjacent surface layers. Unfortunately, the interpretation of diffraction intensities is highly complicated by multiple
diffraction, and kinematic or first Born approximation is not valid as in X-ray diffraction. In most cases even the extension to higher Born approximations does not
lead to convergence, and it was only in the early 1970's that suitable numerical
methods with tolerable computer effort were developed, becoming improved in the
following years (for comprehensive reviews see /1.31,32/). So the recent increase
in structural determinations as shown in Fig.1.1 results from computational progress. The peak occurring at about 1977 is believed to be at least partially due
to repeated structure analyses and therefore should be reduced to some extent
to give a more reasonable increase in activity. Up to now, more than a hundred
structural determinations have been performed for clean and adsorbate-covered surfaces (for reviews see /1.33,34/). Today computer programs are available which can
be applied also by those experimentalists who do not wish to consider theoretical
details. However, only comparatively simple systems can be treated successfully and
they still necessitate considerably large amounts of computer time, allowing only a
few structural or nonstructural parameters to be varied.

Nevertheless, a great deal of surface models with an unit cell area below about
25 $\mathring{A}^2$ and containing not more than 4 atoms can be calculated yielding an accuracy
of atomic positions of about 0.1 $\mathring{A}$. Even so, this or determinations of even higher

accuracy can only be obtained if sufficiently accurate measurements are available for comparison. In contrast to the situation a few years ago, theory has reached a stage where the improvement of experimental reliability becomes important. It is the purpose of this paper to report the most recent developments in data collection and data handling in LEED.

We start the next chapter with a description of the current standard of theory-experiment fit. It will be demonstrated in more detail why more reliable measurements are necessary. In Chap.3 several new methods for LEED intensity measurements are reported which promise more accuracy by fast and convenient measurements. The following chapter discusses results obtained with the new methods as compared to the usual measurements of intensity spectra, and the last chapter concentrates on new possibilities which are feasible by the new methods.

2. Structural Determination by Comparison of Experimental and Calculated Intensities

Experimental data for the determination of surface structures by LEED come from the intensities of different diffraction spots recorded as a function of energy or angle of incidence of the primary electrons impinging on the sample. So sets of energy-intensity spectra $I(E)$ or rotation diagrams $I(\Theta,\Phi)$ are usually collected. In some cases also so-called iso-intensity maps are presented /2.1,2/ which contain intensity contour lines in a plane whose axes are defined by the energy and the polar angle of incidence. The spectra have to be matched by model calculations because no direct method for model determination exists. There is no evident reason why one of the methods, energy or angle variation, should be favored. $I(E)$ measurements can be generated more easily because only a voltage has to be varied, while $I(\Theta,\Phi)$ curves at constant energy have the advantage that energy dependences of the scattering potential and of the optical potential do not enter into the calculation. In both cases, however, the problem still remains of how to properly compare calculated and experimental spectra with detailed features in order to develop the surface model which best approximates reality.

First, a measurement of the agreement between two spectra must be defined since comparison by simple inspection is too inaccurate. We will stress this point in the following section and demonstrate the necessity for precise measurements. In Sect.2.2 measurements of different laboratories on the same surface structure will be compared. It will be shown that the agreement between different experimental spectra

4

of the same surface can be much worse than the best fit of experimental and calculated spectra reached by variation of surface-structure parameters. As the great
majority of existing calculated or experimental spectra are I(E) curves, we will
concentrate on that class of data.

2.1 Comparison of Theoretical and Experimental Spectra

The simplest description of the difference of two sets of data by a least mean
square fit does not consider the peak structure of intensity spectra. More suitable
criteria, e.g., the average deviation of peak positions, have been tried. Though
this measure was successful, at least in some cases /2.3-5/, it has the disadvantage of neglecting the relative intensities of different peaks as well as their
widths. Other more-or-less simple criteria have been tried, also with limited success /2,5.7/. However, a proposal of ZANAZZI and JONA /2.8/ is more sophisticated.
The so-called reliability or r-factor is defined in order to emphasize the structure of the spectra rather than their peak heights. Therefore, the intensity spectra to be compared, $I_1(E)$ and $I_2(E)$, are normalized with respect to their average
values, $<I_1>$ and $<I_2>$, and only the differences of the derivatives, $I_1' = dI_1/dE$ and
$I_2' = dI_2/dE$, are considered:

$$r = \frac{1}{\Delta E} \int_{\Delta E} W(E) \left| \frac{I_2'}{<I_2>} - \frac{I_1'}{<I_1>} \right| dE.$$

The integral is taken over the energy range ΔE which defines the range of overlap
between the spectra, and W(E) is a factor which additionaly increases the weight of
particularly narrow maxima using the second deriatives I_1'' and I_2'',

$$W(E) = \frac{\left| \frac{I_2''}{<I_2>} - \frac{I_1''}{<I_1>} \right|}{\left| \frac{I_1'}{<I_1>} + \varepsilon \right|}.$$

In order to avoid a divergence of W(E) at the position of an intensity maximum, ε
has to be different from zero. It was shown to give best results for
$\varepsilon = |(I_1')_{max}|/<I_1>$, were $(I_1')_{max}$ is the maximum value of the first derivative in
the energy range under consideration. In the experiment-theory fit the indices 1
and 2 denote experimental and calculated spectra, respectively.

Two arbitrarily different spectra reveal an r-factor of $r_{arb} \sim 0.027$. Therefore, the comparison of curves which are to be fitted to each other should be normalized with respect to this value and the so-called reduced r-factor results in $r_r = r/0.027$, which is sometimes used synonymously with the expression "r-factor". It must be kept in mind that $r_r = 1$ is no upper limit for r_r, as two spectra may differ systematically and not only statistically.

As an example, Fig.2.1 presents the comparison of calculated and measured spectra of the 1/2 1/2 diffraction spot of Ni(100)-c(2×2)CO taken from /2.9/. This adsorbate system was the object of a number of investigations resulting in controversial results. However, we will stress this point in more detail in Chap.5 and concentrate at the moment on Fig.2.1 only. In the calculated spectra, the bonding lenghts of Ni-C and C-O corresponding to a vertically oriented CO molecule on top of an Ni atom were varied. Only the curves yielding best fits are shown together with the measurements. In Fig.2.2 an r-factor map is given, which shows contour lines for the r-factor as a function of the C-O and Ni-C bond lengths. It appears that a minimum $r_r \sim 0.1$ results for values of about $d_{CO} \sim 1.15$ Å and $d_{NiC} \sim 1.8$ Å with a precision of about ± 0.05 Å. Of course, the structure determination should be based not only on one beam but on the data of as many beams as possible. The r-factors for the corresponding beams add up, weighted by their respective energy ranges ΔE_i, and result in a mean value $\bar{r}_r = \sum \Delta E_i (r_r)_i / \sum \Delta E_i$. As the reliability of a structure determination should increase with the number of evaluated beams, an additional weight was introduced by ZANAZZI and JONA to give the "structure R-factor" $R = (3/2n + 2/3)\bar{r}_r$ with n the number of beams used. Of course, this definition is arbitrarily to some extent and was chosen in /2.8/ to urge the application of more than 4 beams. Evaluation of more than 10 beams, however, yields no further reduction. In the example of Ni(100)-c(2×2)CO including also the 10 and 11 beam spectra, a minimum was found in /2.9/ to be $\bar{r}_r = 0.19$ (R = 0.22) for the bonding-length values given above. However, larger error bars result giving $d_{CO} \sim (1.15 \pm 0.1)$Å and $d_{NiC} \sim (1.8 \pm 0.1)$Å. This is in agreement with recent results of other authors /2.10-12/ who favor values $d_{CO} \sim (1.15 \pm 0.1)$Å and $d_{NiC} = (1.7 \pm 0.1)$Å.

It must be emphasized that the reliability factor defined as described above is not necessarily an optimal measure in all cases. So, for example, if only one strong but narrow experimental peak is not reproduced by the calculation, a poor r-factor results even with otherwise perfect agreement. Most recently another r-factor construction without using derivatives has been proposed /2.13/ which gives similar results. A more detailed and most recent discussion on r-factors can be found in /2.14/.

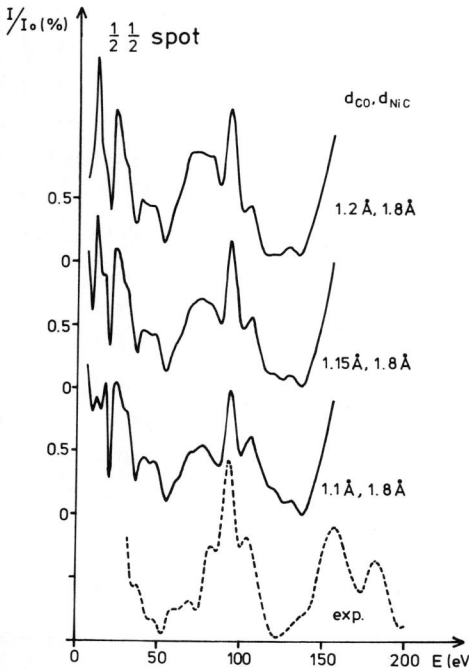

Fig.2.1. Experimental and calculated spectra of the 1/2 1/2 spot of Ni(100)-c(2×2)CO
for normal incidence of the primary beam
and T ~ 100 K taken from /2.9/. Calculations were varied with respect to the
bond lengths of C-O and Ni-C assuming
top sites on Ni for CO molecules with
vertically oriented straight axes

Fig.2.2. Map of the reduced r-factor for
the 1/2 1/2 spot of Ni(100)-c(2×2)CO as
a function of the bond lengths Ni-C and
C-O

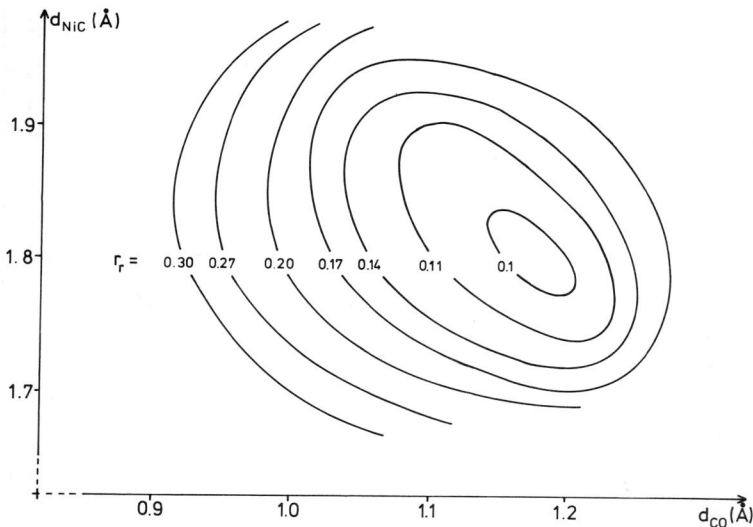

It must therefore be concluded that in spite of some apparent disadvantages, the
r-factors in use are the only impartial measures which, at present, are available
for the comparison of rather structured spectra.

Experience has shown that a r-factor value below 0.1 for the theory-experiment comparison of a single spectrum stands for nearly perfect agreement. The lowest value which has come to the authors' knowledge for an one-beam comparison is r_r = 0.04 /2.8/, which however, was based on a relatively short-ranged spectrum with poor structure. A recent structure determination on Cu(100) gave a comparatively low value of R = 0.068 in a four-beam analysis /2.15/. The average quantities $\bar{r}_r$ or R are often larger than 0.1 for the best fit, and it has been proposed /2.8/ that for R $\lesssim$ 0.2 the structure model under consideration can be taken as very probable, for values of about R ~ 0.35 as probable and for larger values as doubtful. Of course, this classification is again to some extent arbitrary, but is supported by experience. As demonstrated above for the example of Ni(100)-c(2×2)CO, bonding lengths are determined with an accuracy of typically ±0.1 Å.

The question arises why the agreement between calculated data and experimental spectra cannot be improved at the moment in order to increase the accuracy of the structure determination. The answer is certainly twofold. First, the calculations are still to some extent uncertain. This is mainly due to the fact that the scattering potential of the atomic scatterers is not known with sufficient accuracy. An impressive demonstration of that influence has been given recently /2.16/ for calculations on the W(100) surface. Another reason comes from the fact that it is impossible to vary *all* surface parameters which are free to be adjusted. The main limitation for improvements, however, is the restricted accuracy of the measured spectra. We will illustrate this point in the following chapter.

2.2 Comparison of Experimental Spectra from Different Measurements

In the paper already cited above /2.16/, LEED intensity measurements of W(100) taken by different authors /2.16-20/ were presented for comparison. This surface is very well suited for a display of experimental intensities since it has been investigated for more than a decade. In the present work some spectra are given once more in Fig.2.3, but normalized with respect to the peak at about 110 eV in order to facilitate visual comparison. Additionally, a measurement taken by the present authors' LEED group is appended /2.21/. As given in the figure caption, most of the curves refer to elevated and slightly different temperatures in order to avoid a mixing of the (1×1) structure with the low temperature c(2×2) phase of W(100).

8

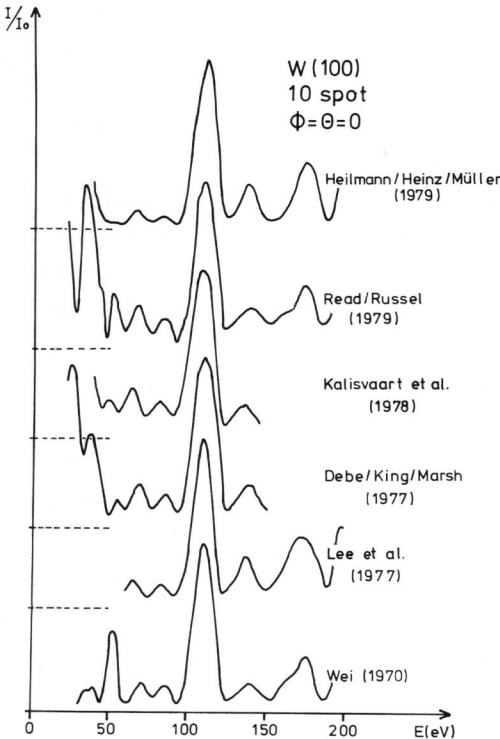

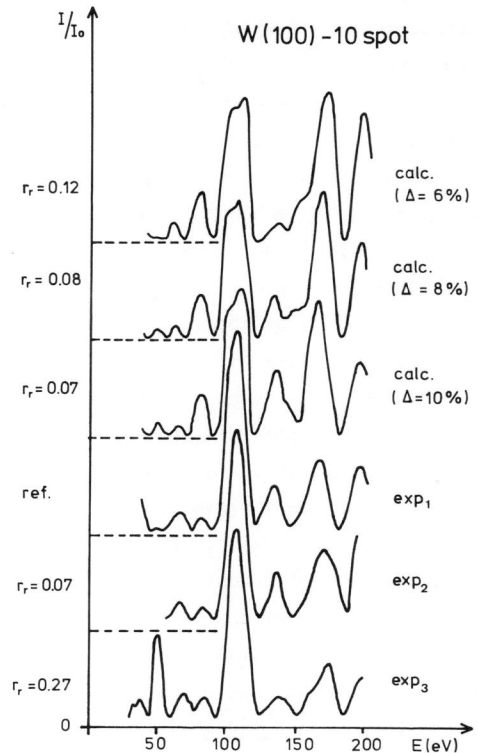

Fig.2.3   Fig.2.4

Fig.2.3. Comparison of different measurements of the 10 spectra of W(100) for normal incidence. The curves are taken at different temperatures, i.e., from top to
bottom at 350 K /2.21/, 650 K /2.16/, 550 K /2.17/, 470 K /2.18/, 300 K /2.19/ and
558 K /2.20/

Fig.2.4. Comparison of several experimental and calculated spectra of the 10 spot
of W(100) for normal incidence. The experimental curve on top of Fig.2.3 is taken
as the reference for r-factor evaluation which is performed for two other experimental curves ($\exp_{1,2}$) from Fig.2.3 as well as for best fit calculations with varying surface contraction $\Delta = \Delta d/d$ taken from /2.21/

However, as it was pointed out in /2.21,22/ the integer spots are only negligib-
ly affected by the reconstruction. Because of the high Debye temperature and high
atomic mass of tungsten, considerable changes with respect to the structure of the
spectra cannot be expected to be caused by elevated temperatures.

It is evident by visual inspection of the spectra in Fig.2.3 that they differ
with respect to relative peak intensities as well as to peak shapes. However, a
comparison based on the r-factor formalism should be performed to show the dis-

agreement on a quantitative scale. Moreover, the extent of disagreement should be contrasted with the r-factor value of the best experiment-theory fit. This is done in Fig.2.4 where the authors' curve (of Fig.2.3 denoted exp_1) is taken as a reference, because r-factor data on its agreement with calculated spectra are available /2.21,22/. It turns out that the best fit, r_r^{calc} = 0.07, occurs for a surface layer contraction of $\Delta d/d$ = 10%. For experiment-experiment comparison, the spectra of best and worst agreement are once more displayed, taken from Fig.2.3, giving values $r_r^{exp} \sim 0.07$ and $r_r^{exp} \sim 0.27$, respectively.

So it must be stated that—at least in the case of the W(100) surface—the agreement between different experimental results is not better, and even worse, than the possible theoretical fit. Though there is no detailed investigation on the quantitative consequence of this fact with respect to the accuracy of surface-structure determinations, it is very probable that the different results for $\Delta d/d$ which have been reported in the literature are at least partly due to experimental uncertainties. The values reported for the surface contraction concluded from LEED intensity spectra are $\Delta d/d$ = $6 \pm 6\%$ /2.23/, $11 \pm 2\%$ /2.19/, $4.4 \pm 3\%$ /2.18/, $5.5 \pm 1.5\%$ /2.24/, $10 \pm 2\%$ /2.21/ and $8 \pm 1.5\%$ /2.25/. Though one must consider that a 1% difference in $\Delta d/d$ refers to an absolute error of $\Delta d \sim 0.016$ Å only, it is evident that some of the results contradict each other even within their given error width. However, it must be emphasized that the usually given error width only means that for model parameters varying within that width, similar agreement between calculated and experimental spectra results. The experimental data and the numerical approximations are believed to be sufficiently precise. This, however, in many cases is an incorrect assumption, and experimental, as well as numerical uncertainties, may contribute considerably to a realistic error width. So, restricting ourselves to the experimental point of view, the reduction of experimental error appears to be an important and necessary demand for progress in accurate structural determinations.

2.3 Sources of Experimental Error

Experimental uncertainties of intensity spectra may arise in two ways. First of all, the disagreement of different measurements can be due to different samples or to different ambient conditions of the same surface or adsorbate. Secondly, inaccurate adjustment of experimental parameters, as well as primary beam interaction with the specimen, can cause false experimental data. The first case occurs when measurements of different laboratories are concerned. Even "identical" means of preparation generally lead to slightly different surface properties. However, LEED experience demonstrates that these differences affect more the absolute peak intensities rather than the structure of the spectra, with the exception of the case of considerable disorder of the first atomic layer (e.g., /2.26/) or in the presence of surface steps within the coherence area (e.g., /2.27-29/). A more detailed review of the influence of defects is given in /2.30-36/. The behavior described finds its theoretical counterpart in the fact that calculated absolute intensities are by a factor of about 2-5 too high compared with the measurement. The calculations are based on the assumption of perfect surface order while considerable surface roughness can be the reality which reduces absolute intensities.

Thus it appears that different, but well prepared samples reveal very similar intensity spectra. A theory-experiment misfit is therefore unlikely to be caused by surface preparation. We now consider those reasons for differences between experimental and theoretical results which arise from imperfect measurements. The major part is due to at least one of three main groups of error sources which are

- imperfect definition of the angle of incidence of the primary electron beam
- missing or unsatisfactory background subtraction
- slow speed of measurement when surface conditions vary with time.

Though there is no systematic investigation of the consequences of misalignment of the crystal with respect to the incident electron beam, it is known from many examples that deviations from a definite direction of incidence of 1° or less can cause severe modifications of intensities. So for Ni(100), intensity changes of up to 30% have been reported in the case of a 0.5^o deviation from normal incidence /2.37,38/. An example of a more recent measurement /2.39/ is given in Fig.2.5 for the 10 beam measurement of Ni(100) for an angular misalignment of 1° compared with normal incidence. It turns out that the spectra change with respect to absolute intensities as well as to their structure in the whole energy range. Using the r-factor measure, it has been demonstrated for example of Cu(100) and Ag(110) that a 0.5^o deviation from normal incidence causes discrepancies of the corresponding spectra of at least $r_r = 0.1$ /2.40/. In the case of Fig.2.5, even higher values must be expected. Angular deviations of the order of 0.5^o, however, must always be anticipated. A residual magnetic field of 10% of that created by the earth's field can cause a deviation of 1° at 75 eV energy. It will be demonstrated in Chap.4 how the corresponding error can be reduced.

The necessity for background subtraction when LEED intensities are measured is commonly accepted because of contributions resulting from thermal diffuse scattered electrons. However, there is some arbitrariness in the background determination, since the diffraction spot area is not sharply limited and the background may vary within the spot. One method of background determination is to measure the back-

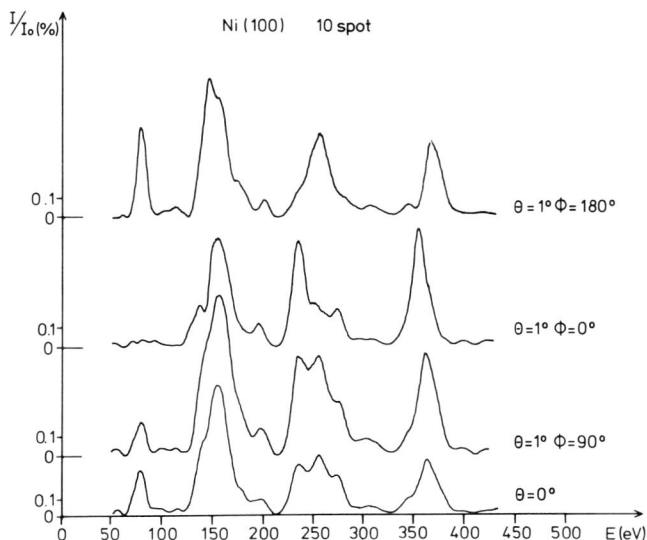

Fig.2.5. Variation of the 10 spectra of Ni(100) at T ~ 300 K for small deviations from normal incidence of the primary beam (Θ = 0). The azimuth Φ = 0 refers to the [10] direction

ground level aside from the spot. Another way is to measure first the signal from a narrow area just containing the diffraction spot and then to perform the measurement in a larger area containing the surrounding background as well. Figure 2.6 demonstrates the effect of background subtraction for the specular beam spectrum of Pt(100)hex /2.41/, which was measured with a spot photometer, but not normalized with respect to the incident beam current. The inserted part of the figure indicates the spot area (hatched) which is practically identical with the area (diameter d) of the first measurement [curve (b)]. The enlarged area (diameter D = 2.67d) results in the spectrum (a). The difference (a)-(b) weighted by $f = (D^2/d^2-1)^{-1} = 0.164$ gives the background of the spot area as presented in (c). The difference (b)-(c) is the background corrected spectrum (d). It is evident that the structure of the spectrum is more pronounced than in the uncorrected curve (b) and that relative intensities of neighboring peaks can change as in the case of the peaks at 225 and 270 eV. Moreover, it shows up clearly that the background spectrum (c) can be structured, i.e., the subtraction of a smooth background from an uncorrected spectrum can lead to inaccurate results. However, also the subtraction procedure described above is based on a spatially constant background

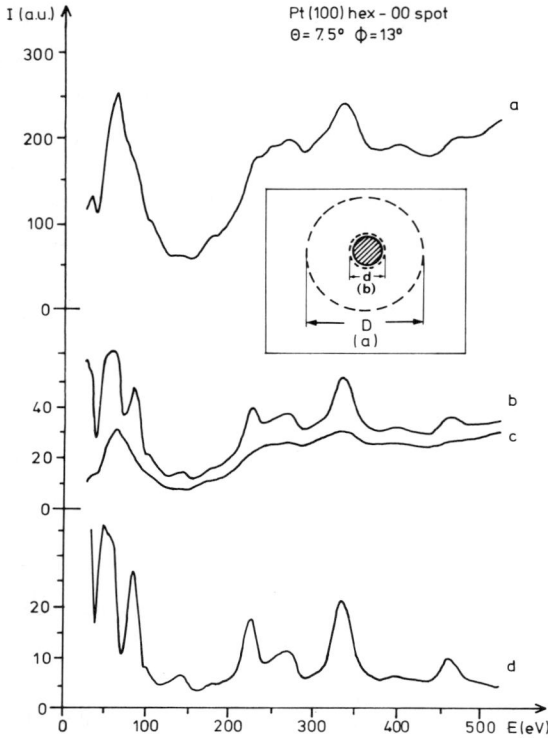

Fig.2.6. Spectra of the specular beam of Pt(100)hex taken from /2.41/ for different apertures (a) and (b) of a spot photometer as demonstrated in the insert where the shaded area represents the diffraction spot. Correction of (b) by the background variation (c) results in the true spectrum (d)

around the diffraction spot and it introduces some error if the background is varying. We will come back to this point in Chap.4.

Besides these experimental errors we must consider structural variations of the surface during the time of measurement. Even for a low residual gas pressure of 10^{-8} Pa, a full monolayer of adsorbed gas can develop on a clean surface during an 1 h measurement when the sticking probability is of the order of unity. In this case the surface must be repeatedly recleaned, and care must be taken to restore the initial surface state. The analogous procedure is even more tedious for adsorbate systems for which the adsorption process must be also repeated after cleaning. However, in these cases much more trouble may be caused by the incident electron beam, which can modify the adsorption structure. The main processes in this sense are electron stimulated desorption of a weakly bound adsorbate (ESD) and/or in cases of molecular-adsorption electron-induced decomposition. Metal adsorbates are mostly unaffected. The total ESD cross section for electrons in the range 50-150 eV is reported to be of the order of 10^{-16}-10^{-20} cm^2 for gas adsorption /2.42/, whereas for Cs/W(100) the value is less than $6 \cdot 10^{-22}$ cm^2 /2.43/ (detailed reviews about

ESD are given in /2.42,44,45/). For an average value of the cross section 10^{-18} cm^2, an impinging electron beam of 1 µA/mm^2, i.e., a current density of 10^{-4} A/cm^2 causes a coverage decrease by a factor of 2 within 20 min. For the case of CO on Ir(111) a cross section of about 10^{-17} cm^2 was found /2.46/ for an electron energy of 86 eV. This corresponds to a similar coverage decrease for a primary beam of only 0.1 µA/mm^2 typical for LEED experiments. But even if molecules adsorbed on a surface are not stimulated by the incident electrons to desorb, they may be caused to dissociate resulting in a modified surface structure. Desorption may then take place after decomposition. A well-known example is the system CO/Ni(100). It was first reported in /2.47/ that for a primary beam current of 1 µA (no value for the current density was given), significant changes in the original c(2×2) superstructure pattern occured after only several minutes of exposure, even for electron energies down to 20 eV. Even more drastic and rapid changes are reported for adsorption systems where organic molecules are involved, these being modified within a very short period of time /2.48/.

Thus, it must be concluded that it is necessary in certain cases to apply a rapid method for LEED intensity measurements. Such a method also facilitates numerous control measurements and can therefore be expected to avoid errors caused by crystal misalignment and/or erroneous background correction. It will be shown in the subsequent section that the demand for fast measurements described above cannot generally be met by the classical methods using a Faraday cup or spot photometer. Therefore, during the last years a number of attempts have been made to develop rapid measuring methods. This will be described in Chap.3, finally concentrating on the most rapid method available today.

2.4 Classical Methods for Intensity Measurements

In principle the LEED intensity determination is very simple, it is the measurement of the steady current of a diffracted beam. The detector, however, has to track the diffraction beam which moves in space when the parameters of incidence, i.e., the angle of incidence or the energy of the primary beam, are varied.

The direct way to measure this current is by a Faraday collector. This is the oldest and original method /1.14,2.49/, and is still the most sensitive one. It has the advantage of giving data on an absolute scale. The sensitivity, at least in principle, is only limited by the leakage current of the cage, which can be reduced to below 10^{-14} A. This is considerably smaller than the current of a diffracted beam when a primary current of about 10^{-7}-10^{-6} A is used. However, the cage has to be carefully shielded in order to repel inelastically scattered electrons and to

avoid secondary electron emission. Long measuring times result since the spectra
are taken point by point, a tedious task. The time to take one intensity point is
typically of the order of 10 s resulting in a total measuring time of 1 h for a
spectrum of 300 points for *one* beam.

Efforts have been made to increase the speed of measurement with a Faraday cup.
One such effort makes use of an automatic system /2.50/, where the collector is
made to move in the plane of movement of the beam under consideration. The collector is moved continuously, while the spot is repeatedly swept over its aperture.
This is done by a corresponding automatic sweep of energy which can be made relatively short ranged in order to avoid successive entrance of different spots of
the same azimuth. The track of the moving cup is followed by a main sweep voltage.
In this way the total measuring time can be considerably reduced, a rate of 200
eV/min was reported /2.50/. However, since the crystal must be azimuthally rotated
for spot selection, a proper crystal holder must be available. The method is restricted to normal incidence of the primary beam when a fixed electron gun is used
as is currently done. This holds also for a modification allowing for a cage fixed
in space /2.51-53/, which has the shape of a slit instead of the classical cup. The
slitlike cage is oriented azimuthally and can be mounted behind a radial slot in
the luminescent screen frequently used in LEED equipment. If the slit coincides
with the track of the spot, very short measuring times are possible. The energy can
be swept continuously and the diffraction-beam current is simultaneously recorded.
The measuring speed is only restricted by RC time constants of the cage, the electron optics, and the sweeping power supply. As a typical time for a 300 eV spectrum,
a total of about 6 s is given in /2.52/. In spite of the enlarged detector area
which causes increased background and noise, a sensitivity of 10^{-14} A could be
achieved using a chopped primary beam and lock-in technique. However, besides the
above described restriction to normal incidence, a further restriction arises since
only one beam at a time should enter the cage. Therefore, the cage should be divided up into segments, but highly complicated patterns could still not be measured.
In this case one must return to the cuplike movable cage with small aperture, typically several millimeters in size, for integral spot measurements. To avoid experimental restrictions, the time-consuming tracking of the spot by hand can only be
speeded up by driving the cage automatically by self-adjustment. This has recently been tried /2.54/ using a movable Faraday cup which follows the spot under computer control by continuous maximization of the signal. The system, which is also
provided with a movable gun, seems to be the most versatile one based on the Faraday cage principle at the moment. However, because mechanical tracking cannot be
avoided, measuring times in the range of 2-4 s per intensity point result, in-

cluding background determination and subtraction as well as auxiliary measurements /2.55/. This, on an average, gives a total of 15 min for a spectrum of 300 points for a single beam.

Some of the technical problems mentioned in connection with the Faraday cup equipment can be avoided by measuring the optical brightness of LEED spots in a display-type LEED system. It works by post-acceleration of the diffracted electron beams /2.56/, where inelastically scattered electrons are repelled by an electrostatic filter. The elastically scattered electrons generate bright diffraction spots with a thermal diffuse background on a spherical luminescent screen. The spot brightness is measured by a spot photometer through a window from outside the UHV chamber. This advantage is gained only at the cost of calibrating the luminescent screen, since only indirect signals can be taken. However, it turns out that the system responds practically linearly to the incident current density of diffracted beams, at least for primary beam currents of not more than 10^{-6} A. Therefore, the measurement of the optical signal results in accurate relative data. This is sufficient for the comparison with theoretical data, because only the structure of the intensity spectra is compared. Also, the absolute diffraction intensities yielded by a Faraday cup are considerably lower than the calculated ones because of unavoidable residual surface disorder or roughness. However, as in most cases the axis of the spot photometer is fixed, the optical transparence of the repeller grid system varies when the diffraction spot moves. Additionally, Lamberts law has to be considered. So a four-grid optics (transparency 0.81 of one grid) has a transparency of about 0.43 when the photometer is directed perpendicular onto the screen as usually verified in the center of the screen. Then the overall response decreases to about 0.21 at the edge of a standard 100° screen when the photometer is shifted parallel to the optical axis. A detailed correction formula has been given in /2.57/ for the transparency of the grids. Though the described variation with the angle of diffraction is pronounced, its influence on the structure of intensity spectra is not very severe because the dependence is smooth and therefore introduces no erroneous structures. In many cases a corresponding data correction is not performed which leads to artificial weakening of low-energy data. This, however, can be limited by measuring the intensities only for higher energies. For the 10 beam of W(100) at 50 eV and normal incidence of the primary beam, a diffraction angle of only about 33° results with an overall response of about 0.33, including Lamberts law which is already near the high-energy limit of 0.43. However, it is important to avoid Moiré structures by proper adjustment of the grids because they may introduce considerable local modifications of the electronic and optical transparency /2.58/.

The sensitivity of the spot photometer is mainly limited by the photomultiplier's dark current and noise. Each diffracted electron impinging on the luminescent screen generates of the order of 10^3 photons, two of which on an average reach the objective of the photometer and one hits the photocathode /2.59/. So with a quantum efficiency near to unity practically every diffracted electron can be detected when the tube is cooled and the collection of stray light can be avoided. The aperture of the photometer can be easily changed in contrast to the Faraday cup, so turn over to spot profile measurements can be easily and rapidly performed.

Though the measurement of spot intensities by a photometer is much simpler with regard to the technical set up of the equipment compared with the Faraday cup method, no considerable reduction of the total measuring time can be achieved. This is mainly due to the fact that the tracking of the moving diffraction spot must be performed by mechanical shift of the heavy photometer. Though it can be facilitated by the use of an image intensifier for tracking control by eye /2.59/, a measuring time of 5-10 s for one intensity energy point leads again to the order of hours for the measurement of the spectra of several beams.

Recently, a more practicable method with respect to spot tracking was proposed using a spot photometer /2.60/. Instead of tracking the spot by the photometer, the spot tracks the photometer which is driven manually and more or less arbitrarily by hand. The proper energy sweep is controlled by two linear motion potentiometers connected mechanically to the photometer. This is certainly faster than the classical method. The authors report /2.60/ that a 300-eV-wide spectrum can be obtained "easily in a few minutes". However, obvious limitations cannot be overlooked. At least in the stage published, the method is confined to normal incidence of the primary beam, because the trace of the spot photometer is yet bound to be radial. Only then a very simple formula results to calculate the electron energy depending on the spot position. Certainly this formula could be extended to nonnormal incidence with some numerical effort, but the path of the photometer is then no longer simply radial. Moreover, the most important uncertainty of the method might occur by nonnegligible deviation of the spot's real trace from its calculated one, e.g., by residual sample misalignment or magnetic fields. However, with some additional refinements the method looks undoubtedly promising.

Another attempt has been made very recently to avoid the cumbersome shifting of the photometer. A system of mirrors directs the light emerging from the spot on the screen into the spot photometer /2.61/. When the spot moves the mirror system is readjusted as to always focus the spot light into the photometer. This readjustment is controlled by computer controlled step motors which make the mirror rotate until maximum output signal is achieved. This method seems to be more precise than the tracking of a calculated trace. However, the speed, which is reported for the system as being 4 s per intensity point /2.62/, may still be too low in some cases.

The primary beam current incident on the sample is generally energy dependent. Therefore, normalization of diffracted beams is necessary no matter which measuring method is used. This is especially important for the low-energy region, i.e., below about 50 eV, where the beam current generated with ordinary electron guns decreases considerably with energy. The signal for normalization can be taken in a separate measurement by directing the primary beam into a Faraday cup or onto the luminescent screen via a negative sample bias. As originally proposed in /2.57/ the procedure can be applied in the case of the spot photometer method to cancel the influence of Lamberts law and the transparency of the grids when the primary beam is deflected to the position of the diffracted beam by an external magnet. However, in order to take advantage of this idea careful magnetic shielding of the electron gun is necessary. Another method for the incident current measurement determines the current drawn from ground into the gun at every energy. In this way the primary beam spectrum can be stored and used for point per point normalization of the diffracted beam data after collection. A modified method takes the primary and diffracted beam current simultaneously and performs normalization by logarithmic subtraction of the corresponding signals /2.63/. In this way normalization is carried out during data collection which cancels fluctuations in the emission of the electron gun.

On a whole it turns out that the measurement of LEED spectra by both Faraday cup and spot photometer is time consuming work. The demand for careful background subtraction as described in Sect.2.3 must also be kept in mind. As a consequence the most sensitive test for proper surface alignment at normal incidence, the comparison of symmetrically equivalent beams, is mostly ignored. However, severe consequences of the long measuring time appear in cases of surfaces changing with time, no matter for what reason. Therefore, in spite of the numerous and successful applications of the Faraday cup and spot photometer methods, during the last years some efforts have been made to develop more rapid and still feasible methods to measure LEED intensities. This will be reviewed in the following chapter.

3. New Experimental Methods for Intensity Data Collection

Nearly all developments of new intensity measurement methods are based on the display type LEED equipment, so no modifications are necessary within the UHV apparatus. This guarantees the applicability of the new methods to nearly all existing LEED systems. They all attempt fast and easy performance in order to overcome the problems outlined in Sect.2.3. Two main groups of methods have been developed so far: those based on photographic methods and those based on the use of a vidicon camera. They are described successively in the following. The fastest method available at present is outlined in more detail.

3.1 Photographic Methods

The photographic methods are in principle very simple indeed. Data collection is done by photographs of the LEED patterns displayed on the luminescent screen. The intensity of single diffraction spots stored on this hard copy is evaluated later by different methods. This procedure, first proposed in 1975 by the authors /3.1/, uses a motor-driven camera and exposure times of typically 1 s. The machine-developed negative image is evaluated with the aid of a microdensitometer which works under computer control. During this procedure each photograph is scanned by the densitometer resulting in a grid of points of variable optical density with a spatial distance of 0.0012 cm or less on the photograph. This two-dimensional density map is stored on magnetic tape. In order to calibrate the film response one part of each film is exposed to a continuous gray edge which transmits light of the same spectrum as emergent from the luminescent screen. From this "sensistrip", a table of densities versus corresponding relative intensities is derived by the densitometer and stored on magnetic tape, too. In the third step of the off-line analysis, the computer searches for diffraction spots on the maps stored. Within a fixed radius around each spot maximum the density spots are converted to intensities. They are added up whenever they exceed a locally determined background level resulting in an intensity point. After evaluation of all frames of the energy sweep, the spectra of all spots are plotted. While the work to be done off-line by the densitometer and computer is rather time consuming, the measurement itself by taking the photographs is comparably quick. For an energy range of 20-200 eV in steps of 2 eV, a total time of 10 min is given in /3.1/, which means an average of about 6-7 s for one intensity point. Comparing again to a spectrum of 300 points, a total of about 1/2 hour results. Though this is comparable to the typical measuring times of Faraday cup and spot photometer methods it must be emphasized that this total refers to the simultaneous measurement of *all beams* visible on the screen. The overall advantage of the photographic method is therefore the considerable saving of time, if a number of beams is to be measured rather than the speed with which one single spectrum can be taken. Moreover, a hard copy of the experimental data is available. But the necessary overall measuring time of at least 10 min up to 1/2 h may be too long in cases of time-varying surface conditions. Furthermore, the off-line evaluation of data demands a considerable amount of computer effort and takes about 20 min per frame. From a practical point of view it is disadvantageous that the actual measurement and the final result, i.e., the intensity spectra, are separated to such an extent that an erroneous measurement cannot be detected and readjusted at once. As a result, the test for normal incidence by comparison of the spectra of equivalent beams is really time consuming, because the result of a readjustment of the sample is only available after film development, densitometer scanning and computer integration.

The dynamic range of the method is mainly limited by the response of the photographic film. In /3.1/ a range of three orders of magnitude was given, which is sufficient in most cases. Very weak spots can be detected by prolonged exposure time, very bright ones by corresponding restriction. This means additional photographs must be taken which increases the total measuring time. Comparison of spectra measured with the photographic method and with a spot photometer showed excellent agreement for Pt(111) /3.1/. The method was also applied to more complicated surface systems such as C_2H_2/Pt(111) and C_2H_4/Pt(111) /3.2/, and to the reconstructed (100) surfaces of iridium, platinum, and gold /3.3/. Similar equipment was applied successfully to the (2×1) superstructure of Si(111) /3.4/.

In order to make the off-line evaluation of the photographs more convenient and efficient, a modified procedure was proposed in 1976 /3.5/. Instead of a microdensitometer, a video camera is used to scan the photographs under computer control. The diffraction spot intended to be measured from a given frame is identified by pointing to its position on the monitor with a joystick, which defines crudely the xy-coordinate of the spot. Then an area centered at this position is digitized. From the resulting profiles the exact position of the spot maximum is determined as well as a preliminary background level definitely outside the spot area. From the signals exceeding this level, the width of the intensity distribution at half maximum is calculated. Then the average of intensities within a thin annulus with diameter of twice that half width is taken as the actual background level. All intensities within a circle of the same diameter are then background corrected, added up to a total intensity, and normalized with respect to the incident beam intensity. A more detailed description of the method including some further improvements and tests has been given in /3.6/. The whole off-line procedure takes less than 30 s for one beam of one photograph, which is considerably less than needed by the microdensitometer method. So, it is concluded that a set of 5 spectra each containing 300 points needs about 12 h. Even for considerably less data points an immediate feedback to the measurement, e.g., for sample readjustment, is not possible. This may complicate the data collection in some cases. Moreover the measurement itself, i.e., the collection of photographs, again takes 10-30 min depending on the number of intensity points required. The method has been successfully applied to surface systems such as Cu(111) /3.5/, Rh(100), Rh(111) and Rh(110) /3.7-9/, Cu(311) and Ni(311) /3.10/, Ge(111) and ZnTe(110) /3.6/, Zr(0001) /3.11/, Rh(100)-(2×2)S /3.12/, Rh(110)-c(2×2)S /3.13/.

The most important advantage of the photographic methods certainly is their capability to store the LEED information of all diffraction spots simultaneously. Relative intensities of different spots are safely measured no matter how many spots show up. So for the first time, intensities of complicated superstructures can be measured, if they are stable within 10-30 min. The disadvantage is the time-consuming off-line evaluation of the photographs, prohibiting the very useful immediate feedback to experiment. There is, however, no reason why the promising use of a vidicon camera for photograph scanning could not be speeded up and extended to the scanning of real time pictures, thus allowing for on-line measurements. The development in this direction took place indeed, however, completely independently of the photographic methods and in the very beginning even slightly earlier. This will be described in the following section. As in most cases television systems work under computer control they will be designated as TV computer methods.

3.2 TV Computer Methods

Although the TV computer methods are all based on a common idea, their technical realization can be considerably different. The video camera takes the diffraction pattern on the luminescent screen from outside the UHV equipment and displays it on a monitor for inspection. The electronic video signal coming from the camera contains the whole information from the pattern. It is given to a computer via a suitable interface which may digitize the incoming analogue signals and perform a preliminary treatment. Because the data come in with a high rate, it depends considerably on the special properties of the interface and computer at what rate the data can be processed and converted into intensities. In principle the primary beam parameters could be varied every half frame, i.e., every 1/50 or 1/60 s (Euro-

pean and American norm, respectively). However, in order to allow for this TV rate
data acquisition, the electronic equipment must be able to extract the integrated
intensity signal for at least one beam out of the rapidly varying video signal.
This corresponds to a linewise scan of the diffraction pattern and demands a very
fast interface and computer. Special hardware and software are necessary, which
are not available in each case. Therefore, most methods developed so far work with
a data acquisition rate lower than the TV rate, i.e., the parameters of incidence
have to be held constant for several or many half frames. These methods are reviewed
first in the following section. At present there is only one method to process data
according to the TV rate which will be described separately.

3.2.1 Data Acquisition Rate Lower than TV Rate

The first on-line TV computer method to measure LEED intensities was published in
1976 /3.14/, an improved version in 1979 /3.15/. The technique can be applied by
using any system consisting of a commercial TV unit, a processing computer and a
time base. The principle features are shown in Fig.3.1. The complete LEED pattern
seen by the TV camera through the window of the UHV apparatus is continuously displayed on the monitor. The video signal passes a low-pass filter which is designed

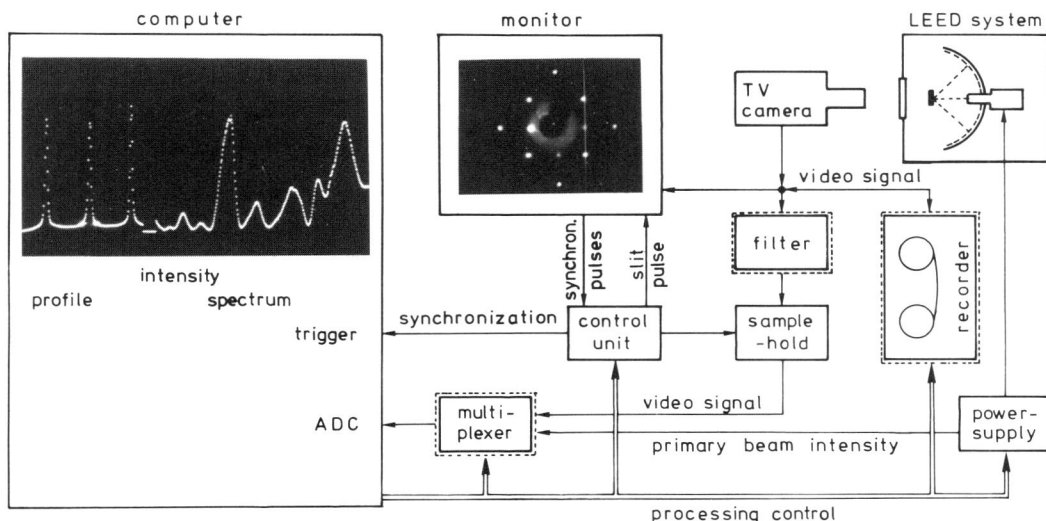

Fig.3.1. Schematic diagram of the computer-controlled TV system taken from /3.15/.
Single lines represent the flow of information, double lines that of processing
control, and elements enclosed by broken lines are optional

to leave spot profiles undisturbed but avoids high-frequency noise. Then it enters a sample-hold amplifier, controlled by an electronic unit which in turn interacts with the processing computer. The sample-hold amplifier is opened at the same chosen position in each TV line and the corresponding video signal is transferred to the computer's ADC via a multiplexer. In this way a vertical cut through the diffraction pattern develops which is displayed horizontally in the left hand side of the computer display. The cut is chosen to hit a diffraction spot and the software is arranged to choose a narrow region of the resulting intensity profile along the line to identify one of several spots touched by the slit. Then the vertical slit operating as an electronic window is shifted over the spot under consideration in small steps of 1/600 of the total frame width. At each slit position the profile of the spot is taken within the time of a half frame. As indicated in Fig.3.2 about

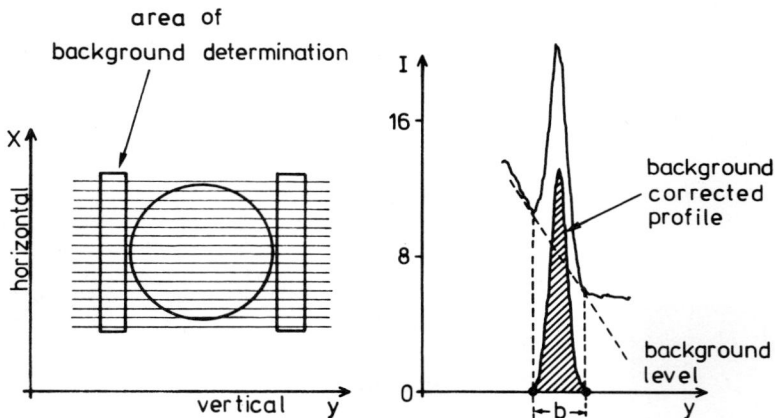

Fig.3.2. Procedure of intensity evaluation by the TV method described in /3.15/. The intensity distribution is digitized along vertical slits (displayed horizontally, left hand side) which are averaged to give a mean profile finally corrected by a linearly varying background

20 slits are necessary to scan the spot completely. The different profiles are added together with simultaneous determination of the spot maximum position and the corresponding half width. The averaged profile contains comparatively low noise and the background level can be safely taken at 2 1/2 times the half width determined. By background subtraction, the corrected mean spot profile results, which is integrated to give the total intensity (Fig.3.2). The input multiplexer of the computer is switched to measure the primary beam current drawn by the power supply

from ground. After normalization the result is stored in the computer memory. Then
the input parameter can be modified, e.g., the energy which is varied by a power
supply programmed by the computer. The energy step width should be small enough
(below 2 eV) to make the new and former spot areas overlap, so the scanning vertical slit can make out easily the new position of the spot maximum and thus follow
the moving spot. If the spot disappears below background level the last maximum
position is held constant until the spot intensity rises again. The intensity spectrum, developing step by step, is displayed on the right-hand side of the computer
display (Fig.3.1) allowing for immediate feedback to the measurement procedure.

The TV computer method described has been proved to be linear in a dynamic range
of three orders of magnitude if the automatic gain control of the camera is turned
off. The spectra measured with this method agree very well with results obtained by
the classical Faraday cup and spot photometer methods, which is demonstrated in
Fig.3.3 comparing measurements of the specular beam of Ni(100) taken by a spot photometer /3.16/ and the described TV method /3.17/. The lowest diffraction beam cur-

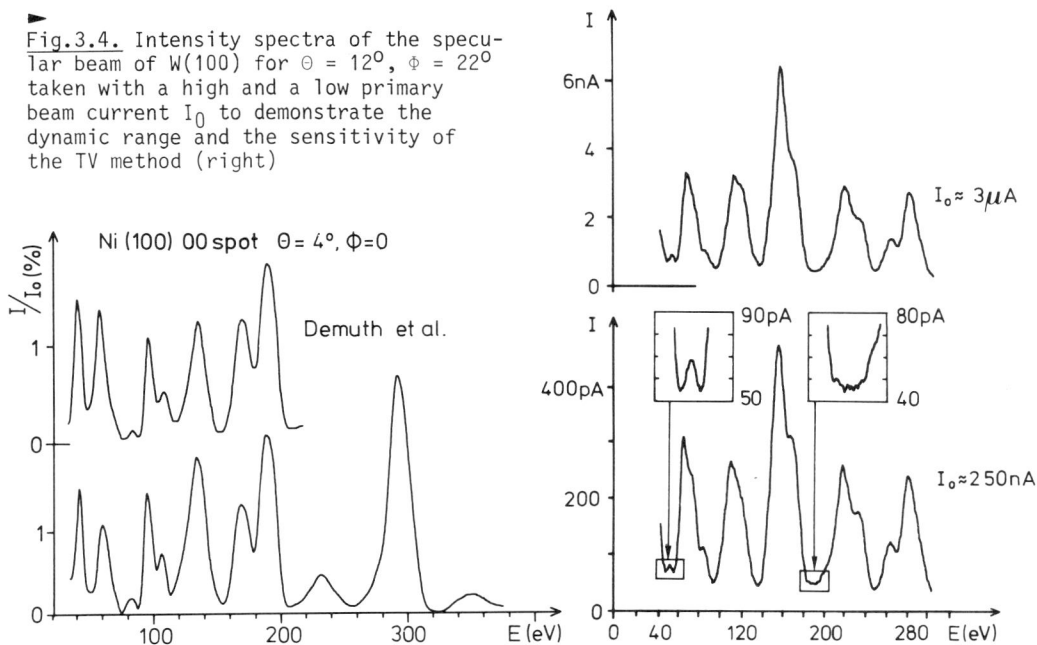

Fig.3.4. Intensity spectra of the specu-
lar beam of W(100) for $\Theta = 12^\circ$, $\Phi = 22^\circ$
taken with a high and a low primary
beam current I_0 to demonstrate the
dynamic range and the sensitivity of
the TV method (right)

Fig.3.3. Comparison of specular beam spectra of Ni(100) /3.17/ measured with the
spot photometer method (lower curve, originally published in /3.16/) and the TV
method (upper curve)

rent detectable was $5 \cdot 10^{-12}$ A. This is demonstrated in Fig.3.4 (taken from /3.14/) for spectra of the specular beam of W(100) which could be transferred to an absolute scale by calibration with a Faraday cup. The upper curve refers to a comparatively high value of the primary current, I_0 = 3 μA, the lower one to a value which is decreased by an order of magnitude. The curves agree very well up to the highest diffraction beam currents of 6 nA. The inset on the left-hand side shows that structures of the spectra reappear undistorted, down to diffraction beam currents in the pA-range, demonstrating the linear dynamic range. The right-hand side inset indicates the noise level of the measurement which turns out to be about 2 pA, so that diffraction beam currents of 5 pA can be detected. These values refer to ZnO luminescent material, 6-keV postacceleration voltage, a 0.7 aperture of the video camera, and a CdSe vidicon camera tube. With higher voltage, more efficient phosphors, or image intensifiers, even higher sensitivity is possible. However, the value attained is sufficient if primary currents of the order of 1 μA are used since even the lowest maxima in intensity spectra do not drop below 10^{-5} of the primary beam current. For extremely high-intensity maxima the camera aperture can be reduced to keep the system working linearly.

Being a real on-line method and the electronic instead of a mechanical tracking of the moving diffraction spots are the most important advantages of the TV computer method described. The speed of measurement is determined by the TV frame rate, which is 50/s (European norm), and by the number of frames needed for the evaluation of data per energy and beam. It turns out that 20 frames are a reasonable value leading to an average measuring time of 0.4 s for one intensity point of one beam. Referring again to a standard spectrum of 300 points, the total of 2 min results which is indeed very small compared with other methods. However, in contradiction to the photographic methods, the spectra of different beams have to be taken consecutively, thus linearly increasing the total measuring time with the number of beams. In the example of 5-10 beams, the time is comparable to that needed by the photographic method. However, the results are immediately available, normalized, and background-corrected spectra can be plotted from the computer memory.

There is a simple way to further reduce the measuring time and to overcome the disadvantage of successive measurement of different beams. As indicated in Fig.3.1, the diffraction patterns can be recorded by video tape prior to off-line computer evaluation. This allows for a higher data acquisition rate and is very similar to the photographic method. In the version reported in /3.15/ the recorder is provided with a single frame display to be triggered externally in order to extract the intensities frame by frame, i.e., energy by energy. The measurement itself can be performed with the half-frame rate, i.e., the intensities of all visible beams are simultaneously recorded within 1/50 s, which means a total of 6 s for a spectrum of

300 points. This is the highest speed of measurement available today which also provides a hard copy as in the case of the photographic method. However, a technically high-standard recorder is necessary and only off-line evaluation of data is possible. This appears to be unfavorable though the time of two minutes for a single-beam off-line evaluation is short compared with the photographic method in the present mode. Moreover, the sensitivity of the TV method is decreased by the intermediate storage of data on the tape. In spite of the use of a high-quality studio recorder, a decrease by a factor of about two was reported in /3.15/. The comparison between the results of the on-line and off-line TV methods, however, shows apparently good agreement as demonstrated in /3.15/. Moreover, the shortcomings of the video recorder use can be overcome when data acquisition with TV rate is possible.. This will be described in Sect.3.2.2.

A way to increase the on-line measurement speed is to reduce the integration area of one spot. So instead of considering 20 vertical spot profiles on an average, only 3 center profiles have been evaluated in /3.15/. This should not be further reduced in order to guarantee the finding of the spot maximum as well as its tracking with the moving spot. The time for measuring one spot intensity is reduced to 60 ms giving a total of 18 s for a spectrum of 300 points and one beam. So for a set of 5-10 beams to be measured, the time, for which the surface conditions have to be constant, is reduced to only a few minutes. The signal-to-noise ratio suffers in this mode by a factor of about two. The most considerable restriction, however, arises from the fact that reliable results can only be attained by this method if the spot shape remains constant when the input parameters are changed during the recording sweep. This condition is certainly not met in all cases.

A method similar to that described above, i.e., using a TV camera and processing computer, is under development /3.18/. A modified method has been published in 1980 /3.19/ not necessarily needing a processing computer. By manual control of two potentiometers, a vertical and horizontal bar can be moved across the TV screen. A simple electronic device takes the maximum intensity in the intersection area of the bars as a measure for the integral spot intensity similar to that described above. The background level is determined in the same way. However, without working in an automatic mode, the measuring time cannot compete with the above method, which also evaluates the spot center only.

In 1979 a different method has been published in two slightly different experimental modifications /3.20,21/. The imaging and digitizing system for the LEED pattern on the luminescent screen consists of a vidicon camera and a commercially available optical multichannel analyzer (OMA) which is connected to a minicomputer via an interface. The vidicon views the LEED pattern as usual. In /3.20/ a

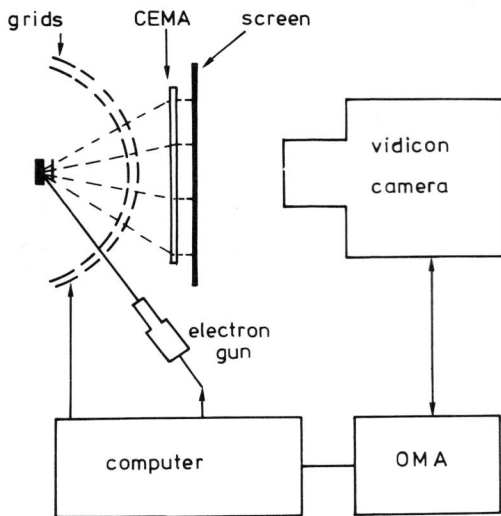

Fig.3.5. Schematic diagram of a computer controlled TV system using a channel electron multiplier array (CEMA) as described in /3.21/. The camera is controlled by an optical multichannel analyzer (OMA) which digitizes the video signal at its channel rate

modification is described which is given in Fig.3.5. The electrons passing the spherical grids are amplified by a channel electron multiplier array (CEMA) which allows a gain of 10^2-10^3 of the diffraction beam currents, as has been already used in /3.22/. The video camera then takes the pattern from a flat luminescent screen arranged immediately behind the CEMA plate which allows reduction of the primary beam current down to the nA-range. The OMA system working in the mode described in /3.20,21/ digitizes the LEED pattern imaged on the video target point by point. Intensity maps, typically 50×50 points, are provided. An example taken in a reduced form from /3.21/ is given for the 7×7 superstructure of Si(111) in Fig.3.6. This digital image can be taken within 1.6 s at each energy which results from the special scanning mode of the video target and the digitizing rate of the OMA. The image is stored on a disk memory unit via the computer, which requires additional time. The energy is stepped forward under computer control and diffraction beam spectra result by off-line computer evaluation of the different digital maps. So for spectra of 300 points, the total measuring time is about 8 min for all visible beams simultaneously. However, a considerable time for data transfer to the disk has to be added and a large disk storage capacity (for 7.5×10^5 intensity points) is necessary. If one restricts data acquisition to beams along a certain azimuthal direction, a corresponding reduction is possible and necessary because of the storage capacity of usual disks. The total measuring time increases if more than one digitizing sweep is necessary to improve the signal-to-noise ratio.

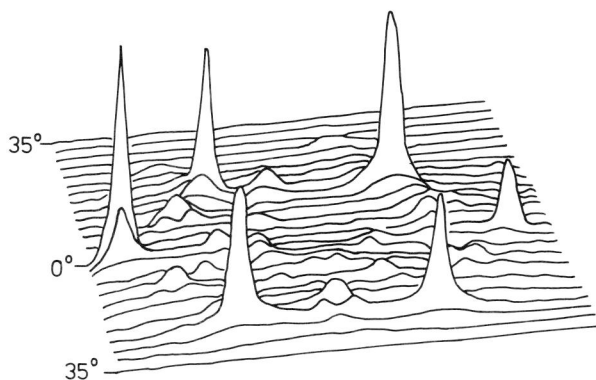

Fig.3.6. Intensity map of the 7×7 surface of Si(111) taken in reduced form from
/3.21/ and measured by the equipment shown in Fig.3.5

Advantages of the method described arise mainly from the fact that complete spatial intensity maps are recorded. Therefore, beam profiles are also available at
any time and the signals of different beams certainly correspond to the same state
of the surface. Half width of diffraction beams from intensity profiles have been
developed in /3.20/ for W(100) and in /3.23/ for GaAs(110). The spatial resolution
is at least 0.75^{o} and can be decreased to 0.07^{o} by reduction of the video target
area to be digitized. The recording of complete intensity maps, moreover, allows
for angle-resolved spectroscopy of photoelectrons, which has been performed in
/3.21,24/. For integral LEED intensities, however, it appears that too much needless digitization is done involving superfluous storage capacity. Only the area
and immediate surroundings of LEED spots are necessary for integral intensity as
well as for profile information. So it seems that the amount of work needed is much
less than that carried out by the method. Moreover, most of the information gained
by digitizing the intensity distribution even just around a spot can be deleted when
the integral spot intensity is extracted. Therefore, data reduction should be done
immediately after digitization. This avoids needless storage of data and comes
closer to a real on-line measurement. The method described in the next section
fits these demands.

3.2.2 Data Acquisition Rate Equal to TV Rate

When a LEED pattern is displayed on a TV monitor, all intensity information is available from one half frame which is written within 1/50 s European standard (1/60 s
American standard). The information usually wanted is the integral intensity of all

visible spots corrected after background subtraction. In many cases the intensity profiles of spots, especially their half widths, are of interest in order to learn something about the degree of surface order. If these data can be taken from the video signal within the half frame time of 1/50 s, the input parameter of the primary beam, as energy or angle of incidence, can be varied with the half frame rate. Intensity spectra could therefore be taken very rapidly indeed. However, to meet this demand, very fast converters from analogous to digital representation (ADC) of data are necessary. They had not been available in the seventies with necessary resolution. Therefore, the above described TV methods were slower than optimally possible. However, very fast ADC's with sufficient resolution were recently developed and made available at reasonable prices. In the device published in /3.25/ and described in the following, a 13.5-MHz ADC of 8-bit resolution was used.

With this fast ADC each of the 290 TV lines (European standard), taking 54 µs/scan plus an additional return time of 10 µs, can be divided up into 730 quadratic intensity pixels. In order to perform a first data reduction, two neighboring data points of each line are added in a 9-bit memory. So the half frame is made up of 365×290 intensity pixels which leave the digitizer unit within 1/50 s. It is obvious from these values that usual processing computers are not able to handle this rate of data input. Therefore, a special interface is necessary which reduces the amount of data to be transferred to the computer comprising some preevaluation.

As demonstrated in Fig.3.7, the video signal from the camera goes to the monitor and via a low-pass filter to the interface which is connected to the computer, monitor and power supply, as well as to a video recorder (optional). The position (x,y) and size (∆x,∆y) of an electronic window is set by the computer and transferred to the interface which performs a corresponding rectangle on the monitor for visual control. It is shifted at the beginning of a measurement to have the diffraction spot under consideration at its center. The size of the window is fixed for the measurement of a spectrum and chosen to be large enough to cover the spot safely for any input parameters. The intensities of the pixels within the electronic window chosen are loaded into a fast memory with an access time of 20 ns and can be transferred to the computer for profile evaluation if desired. However, in most cases only the integral spot intensity is of interest. Therefore, along with the transfer to the fast memory, the intensities of each line within the electronic window are simultaneously summed up by a fast adder which is also part of the interface. This hardware summation, which is demonstrated schematically in Fig.3.8, is carried out for each line of the electronic window separately. Also, the maximum intensity position within each line is determined to allow for later spot tracking when the spot moves. The computer finally develops from the sum of each

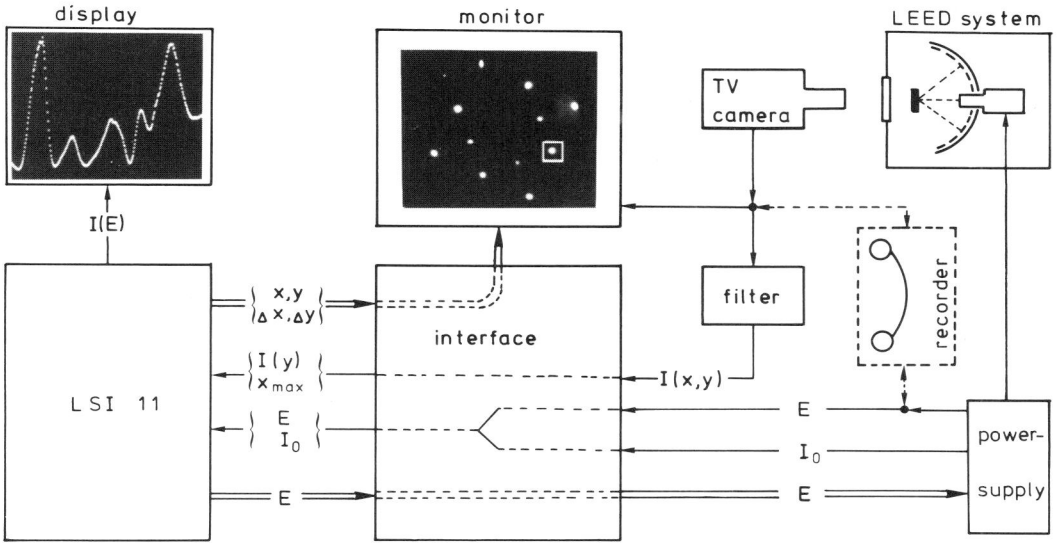

Fig.3.7. Schematic diagram of the computer-controlled TV system with TV rate data acquisition and on-line evaluation described in /3.25/. Simple lines represent the flow of information and double lines that of processing control. A special interface allows for suitable data reduction which makes the measurement of a spectrum of 300 points possible within 6 s

line a line-integrated profile of the diffraction spot (Fig.3.8, upper right). This profile then is background corrected, the background level being determined at the two vertical edges of the electronic window. Since the two background levels determined may differ, an effective background is subtracted which varies linearly between these margins. Finally the peak position of the corrected profile (Fig.3.8, lower right) is determined and stored, the profile is software integrated, and the result is loaded as one intensity point into the computer memory (Fig.3.8, lower left).

After the background corrected intensity has been determined, the input parameters are measured, e.g., the electron energy as indicated in Fig.3.7. This might be important in order to avoid errors by delay effects if very fast sweeps are performed. Immediately after the energy has been measured, the corresponding current of the primary beam is determined and used to normalize the measured diffraction intensity. Then the computer steps the power supply forward and the spot moves correspondingly. The electronic window is still centered around the maximum position determined for the last energy. Therefore, the energy step width must be chosen small enough to guarantee the electronic window to still contain the spot in its

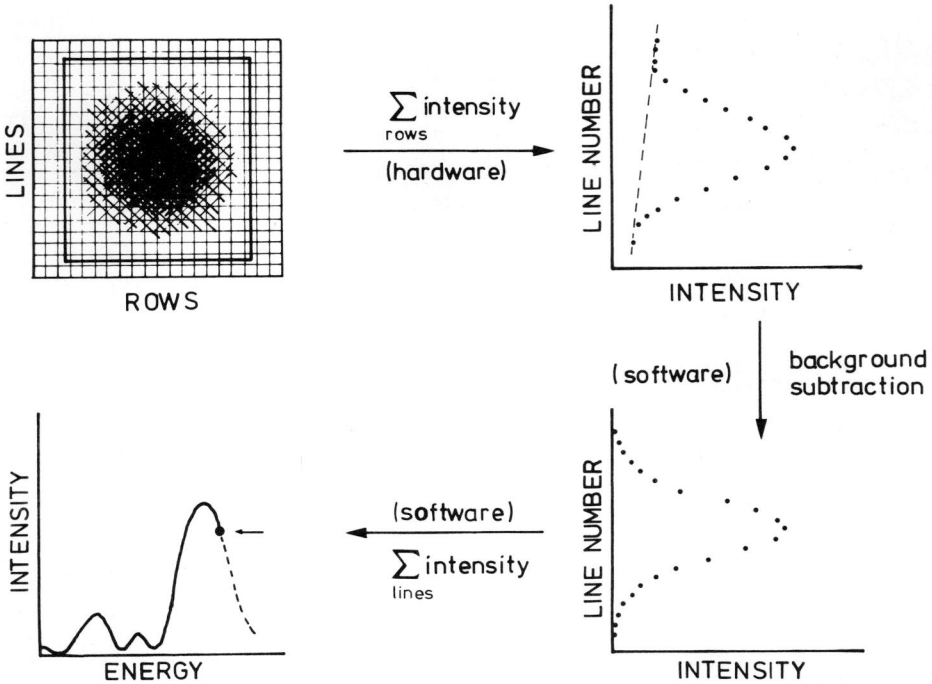

<u>Fig.3.8.</u> Diagram of the important steps during the measurement of one intensity energy point by the fast TV method of Fig.3.7. The whole procedure takes 1/50 s

new position. Values not exceeding 1 eV have proven to be very suitable. The procedure for intensity determination now is repeated including the determination of the new position of the intensity maximum which is used as the center of the electronic window in the next energy step. Thus the moving spot is tracked with an offset of one energy point. If a spot disappears below background level the window position is fixed. If the spot reappears within the window self adjustment of the window position takes place. In other cases the operator has to reset by hand.

The whole procedure of digitization and evaluation of the data within the electronic window resulting in the integral spot intensity is executed within the half frame time, i.e., within 1/50 s. So, a spectrum of 300 points develops extremely rapidly within 6 s, which is the fastest measurement achieved so far. However, this is for only one beam and the measurement of several beams linearly increases the total time needed. It is possible to preselect at the beginning of the measurement up to 4 beams to which the electronic window jumps consecutively for one fixed energy value. In this way the spectra of four beams are measured within 24 s, including background subtraction. An extension to many beams is of course possible. Besides, a repetition of the measurement for another set of beams can be performed as well.

If the total information of the electronic window is needed, e.g., for profile evaluation, the fast memory of the interface has to be transferred to the computer. This again needs the time of a half frame, but only doubles the measuring time. However, as one electronic window consists of about 20×20 pixels, a large amount of data accumulates if no further reduction is performed.

When very large numbers of diffraction spots have to be measured, e.g., for a complicated superstructure and high energy, it might become reasonable to store the analogous intensity data, i.e., the complete diffraction pattern, on magnetic tape prior to evaluation. The information wanted can then be extracted off-line in the same way as described. The only difference is that the electronic signal comes from the recorder instead from the video camera. However, in contrast to the video recorder use described in Sect.3.2.1, no single-frame display recorder is necessary. This is due to the possibility of TV rate data acquisition which can be performed from the running video tape. So, the signal-to-noise ratio no longer suffers from the single-frame display, and intensity data can be taken with nearly the same sensitivity as in the on-line mode. Moreover, it turns out that a normal quality recorder is sufficient.

Undoubtedly the greatest advantages of the method described are its high speed and its on-line data evaluation. The sensitivity and dynamic range as well as other properties are identical to those of the original TV method published in 1976 and described in Sect.3.2.1. Intensity measurements are performed almost by pressing a button, and the normalized, background-corrected spectra are shown on a display or read out by a plotter. The method makes measurements of control easy enough to be performed that they really will be done. For the same reason, more complete data sets will be measured. Moreover, a hard copy of the video data can be saved in analogous form on magnetic tape. In this way the advantage of the photographic method is met. A more complete comparison of the different existing methods is given in Table 3.1.

Applications of the new methods will be demonstrated in the following. We concentrate in Chap.4 on measurements which would be possible with the classical methods, too, but can be performed now with considerably higher precision and reliability. In Chap.5 we turn to new kinds of intensity measurements which are only possible with high-speed equipment. Further applications of the method described in this section can be found in /2.39/.

Table 3.1. List of different intensity measurement methods existing to date. The data correspond to systems as they are actually verified and described in the references cited. The sensitivity data of the TV methods correspond in one case (++) to the use of a channel electron multiplier array and in the other cases (+) to the use of a CdSe vidicon camera.

In the latter cases the use of an intensifier camera would improve the sensitivity by at least one order of magnitude. The sensitivity of the video tape mode can be also considerably improved when no single frame display is necessary for evaluation /3.25/

	METHOD	Time for spectra of 300 points — 1 beam	4 beams	Advantages/ Disadvantages	Sensitivity	Dynamic Range
DIRECT — FARADAY COLLECTOR	cup tracks spot /2.46,47/	~ 1 h	~ 4 h	only normal incidence, azimuthally rotatable crystal holder necessary	< 10⁻¹⁴ A	beyond practical relevance
	moving cup and sweeping spot /2.48/	order of minutes	~ 10 min (+ readjustment)			
	slit like cup /2.49-51/	~ 6 s	0.5 min (+ readjustment)			
	cup tracks spot under computer control /2.52,53/	~ 15 min	~ 1 h			
SPOT PHOTOMETER	photometer tracks spot	~ 1 h	~ 4 h	only normal incidence, no control of exact spot tracking	< 10⁻¹⁴ A	beyond practical relevance
	spot tracks moving photometer /2.58/	order of minutes	~ 20 min (+ readjustment)			
	tracking by rotatable mirrors /2.59/	~ 20 min	~ 1.5 h			
INDIRECT — PHOTOGRAPHIC METHODS	evaluation by computer controlled { microdensitometer /3.1/ ; TV camera /3.5/ }	≤ 30 min	≤ 30 min	all spots simultaneously, hard copy, profile extraction possible, background correction, off-line evaluation	exposure time dependent	≧ 3 orders of magnitude
TV COMPUTER METHODS	On-line evaluation slower than TV rate /3.14,15/	~ 2 min	~ 10 min	automatic normalization	~ 5·10⁻¹² A	~ 3 orders of magnitude
	TV rate recording on tape, off-line evaluation /3.15,25/	6 s	6 s	all spots simultaneously, hard copy	~ 10⁻¹¹ A (+)	
	OMA, intermediate disk storage, off-line evaluation /3.20,21/	8 min	~ 8 min	background correction, profile extraction possible	< 10⁻¹⁴ A (++)	
	on-line evaluation with TV rate /3.25/	6 s	24 s	automatic normalization	~ 5·10⁻¹² A (+)	

4. Examples for Reliable Intensity Data Obtained by the New Methods

The most common sources for errors in LEED intensity data are sample misalignment, missing or insufficient background correction, residual gas adsorption during long-time measurements, and decomposition or desorption of adsorbates by the primary electron beam. It will be demonstrated in the following that their influence can be considerably reduced using high-speed and easy-to-handle methods.

4.1   Influence of Sample Misalignment

Many measurements of LEED intensities are carried out for normal incidence of the primary beam because dynamic calculations of spectra to which they have to be compared for testing surface models can be performed with much less computational efforts by taking advantage of the resulting symmetry. However, as demonstrated already in Fig.2.5, even small deviations from normal incidence, such as 1^{O} or less, cause the spectra to change considerably with respect to relative peak heights as well as to peak positions. This may be true also for deviations from other angles of incidence, but is less investigated for this case. Moreover, for normal incidence an excellent test for misalignment can be performed by critical comparison of the spectra of symmetrically equivalent beams. This test is most sensitive. From an excellent agreement of equivalent spectra, it can be concluded that residual misalignment is of no importance. However, if no perfect agreement can be achieved by careful adjustment of the sample, the degree of disagreement is a measure of the uncertainty or reliability of the measurement. It can be transferred to a quantitative scale by a r-factor comparison of equivalent beams. This is an important point, but it must be emphasized that systematic errors of the measurement cancel by this procedure.

In Fig.4.1 an example is given for the four equivalent beams of the system Ni(100)-c(2×2)CO /2.7/. Their spectra compare favorably well except for the lowest energy peak height which indicates the influence of a residual magnetic field. So this peak should not be considered in theory-experiment comparisons. However, in general the influence of a misalignment cannot be cancelled by simple data truncation, but there is no other remedy of the problem in cases where the misalignment is not too severe. Though there is no theoretical evidence, it seems reasonable to assume that spectra averaged with respect to equivalent beams approximate the normal incidence data, i.e., the averaged spectrum should be much closer to

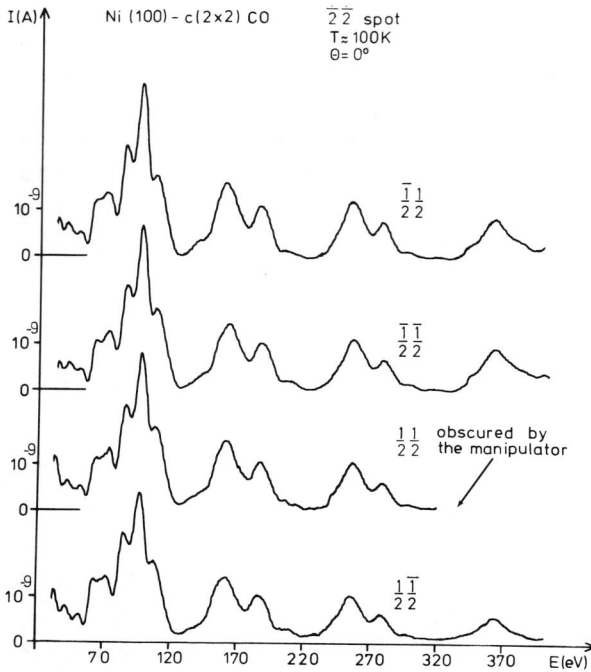

Fig.4.1. Comparison of equivalent half order spots of Ni(100)-c(2×2)CO for best adjustment of normal incidence of the primary beam

the ideal spectrum than any of the single beam spectra which have been recorded under misalignment. This is demonstrated in Fig.4.2 for the 10 beams of clean Ni(100), whose sensitivity to off-normal deviation was demonstrated in Fig.2.5. While the upper curve shows the practically ideal normal incidence, the following ones correspond to angular deviations in 1° steps and represent the average of the four beams 10, 01, $\bar{1}$0, 0$\bar{1}$. It is obvious that the average for $\Theta = 1^o$ is still very near to the ideal spectrum, at least closer than any of the single beam spectra of Fig. 2.5. Also the averaged curves up to $\Theta = 3^o$ agree fairly well, and it is only for $\Theta \geq 4^o$ that considerable modifications take place, especially in the 150 eV region. So it can be expected that averaging of equivalent beams leads to much more reliable data and thus should be done. With a rapid method for intensity measurements this is not unreasonable. The behavior observed can be interpreted by the assumption that the intensities can be linearized as a function of small misalignment angles. The different beam intensities vary in different directions and their changes cancel to the approximation shown. It must be emphasized that the improve-

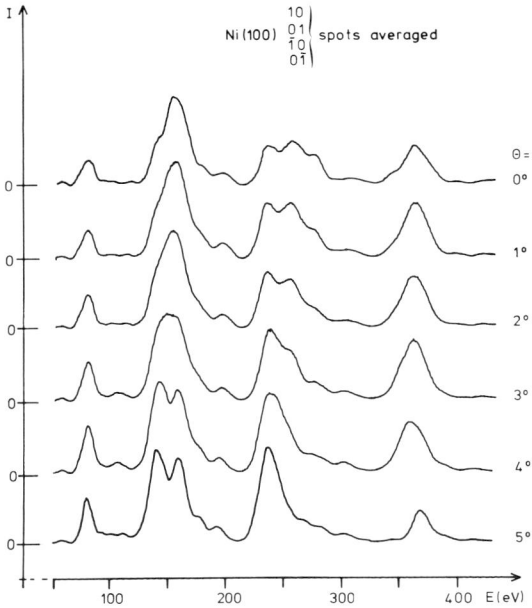

Fig.4.2. Variation of the average spectra of the four 10 spots of Ni(100) equivalent at normal incidence with the polar angle of incidence. Due to the averaging,
considerable consequences of misalignment appear only for a polar angle exceeding 3°

ment of spectra by equivalent beam averaging is a strong argument to prefer data
taken at normal incidence.

The above procedure is of course not useful if information with respect to the
surface symmetry is to be taken from diffraction intensities. Then, extreme care
must be taken for sample adjustment. An example is the clean W(100)-c(2×2) which
shows the superstructure below about 370 K /2.21,22,4.1/, however, clearly developed only for low temperatures. It was observed in /4.1/ that the spectra of different half order beams must be divided up into two different groups. From this it
was concluded that no fourfold but a p2mg symmetry applies to the reconstructed
surface. However, as demonstrated in Fig.4.3 taken from /2.22/, a similar grouping
of spectra of the 3/2 5/2 spots can be achieved by a slight 1° rotation of the
sample with respect to [$\bar{1}$1] axis, whereby normal incidence was adjusted by comparison of equivalent spectra of integer order beams. This does not necessarily dis-
prove the arguments given for the p2mg symmetry, but once more demonstrates the
sensitivity of normal incidence deviations.

I/I_0 (%)

W(100)c(2×2)

$0° < \theta < 0.3°$

$0.5° < \theta < 1°$

0.005

0.003

0.001

$\frac{3}{2}\ \frac{5}{2}$

$\frac{5}{2}\ \frac{3}{2}$

$\frac{3}{2}\ \frac{5}{2}$

$\frac{3}{2}\ \frac{\bar{5}}{2}$

$\frac{5}{2}\ \frac{3}{2}$

$\frac{\bar{5}}{2}\ \frac{3}{2}$

200   300   E(eV) 200   300   E(eV)

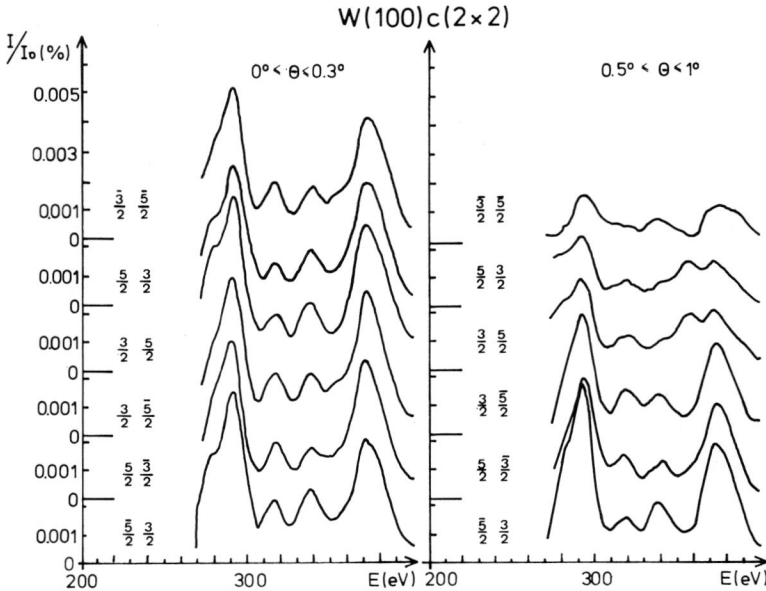

Fig.4.3. Consequence of deviations from normal incidence for some half order beams of W(100)-c(2×2) /2.22/. The curves on the left-hand side correspond to almost perfect normal incidence of the primary beam. A small rotation with respect to the [1Ī] axis by $0.5° < \theta < 1°$ ($\Phi = 45°$) makes one group of spectra change considerably while the other is nearly unmodified (right-hand side)

4.2   Influence of Background Subtraction

Especially at temperatures above the Debye temperature and large momentum transfer, i.e., at high energies, the background subtraction is of considerable importance. In cases of statistical disorder in the surface, e.g., for nonideally ordered adsorbates, this is even more pronounced. With the classical methods the demand for background determination at least doubles the time of measurement as the background level has to be measured separately aside from the spot. In the TV method described in Sect.3.2.2 the intensity measurement is performed by digitizing the intensity distribution within an area containing the diffraction spot, so the background level is determined anyway and no additional work has to be done. Even background spectra could be easily obtained if theoretical progress would allow extraction of definite information from them. The importance of background correction is demonstrated in Fig.4.4 for the adsorbate system Ni(100)-p(2×2)O for the spectrum of the lowest half order beam taken from /3.17/. The upper curve refers to a spot photometer measurement taken from /4.2/ followed by a curve obtained by the TV method. The two

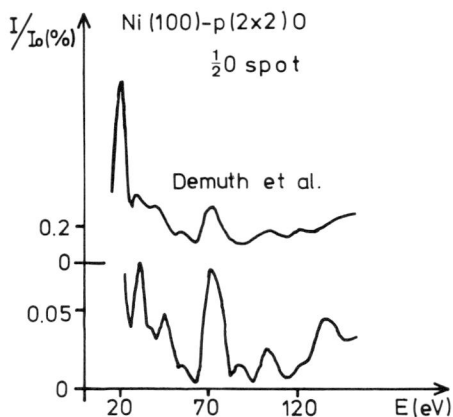

Fig.4.4. Comparison of I(E) spectra which have been measured using background subtraction (lower curve /3.17/) and without correction (upper curve /4.2/)

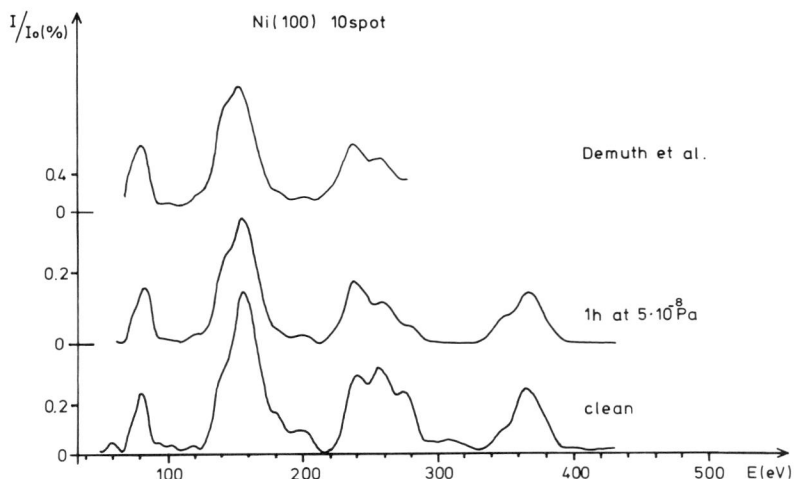

Fig.4.5. Dependence of the 10 spectrum of Ni(100) under residual gas adsorption at room temperature. The lower curve corresponds to a measurement which was performed within a few seconds immediately after a flash so that the surface is believed to be clean. The second spectrum is taken after a 1 h residual gas adsorption /3.17/. Relative intensities in the 250 eV peak group appear to have changed similar to the result to a slow spot photometer measurement (upper curve, /3.16/)

curves show considerable discrepancies presumably due to missing background subtraction in the upper curve. The structure of the uncorrected spectrum is much less pronounced, especially in the region of higher energies.

4.3   Influence of Residual Gas Adsorption

Residual gas adsorption can become important even at very low pressure if long
measuring times have to be accepted. At a partial pressure of 10^{-8} Pa, oxygen adsorption of 0.1 monolayer takes place within a quarter of an hour for a sticking
coefficient of unity. The corresponding influence on intensity spectra may not be
severe, i.e., peak positions remain unaffected. However, Fig.4.5 demonstrates for
the 10 beam of the clean Ni(100) surface that changes of relative intensities of
neighboring peaks may occur. The upper curve refers again to a spot photometer
measurement /2.38/ and should be first compared with the third curve which is the
result of a TV measurement /3.17/. The relative heights of the peak group at about
250 eV are modified. To make sure that this is caused by residual gas adsorption,
the TV measurement was repeated after the sample had been exposed to a residual
gas pressure of 5×10^{-8} Pa for one hour. The corresponding spectrum in Fig.4.5 com-
pares well to the photometer measurement. For further demonstration the results of
the slow and rapid method should be compared for the zero order beam which can be
quickly measured by any method because it is fixed in space during energy sweep.
Figure 3.3 shows that there is almost perfect agreement.

4.4   Influence of Adsorbate Decomposition and Desorption

Adsorbates weakly bound to a surface can be caused to desorb or, in cases of molecular adsorption, to decompose when the primary electron beam impinges onto the
surface system. The only way to keep the amount of desorbed or decomposed adsorbates negligible is either to reduce the primary beam current or the time of measurement. In both cases the total number of electrons impinging on the surface can be
kept small enough to leave the surface system practically unmodified. The case
of primary beam current reduction to the order of 10^{-9} A results in the diffraction beam currents being correspondingly reduced to the region of 10^{-14} A. This
either requires the use of the sensitive classical methods (Faraday cup or spot
photometer) which, however, involve long time procedures of measurement as described
in Sect.2.4, or the use of one of the new handy methods with the application of a
channel plate or an intensifier camera. If, however, a fast method can be applied,
corresponding to the second way cited, additional equipment can be avoided and all
the advantages of an on-line measurement are saved. Nevertheless, both methods are
successful in yielding reliable results.

The sensitivity of an adsorbate system with respect to the incident electron beam can be impressively demonstrated for the system Ni(100)-c(2×2)CO. This is long known to be influenced by primary beam adsorbate decomposition /2.47/. The first structure determination by LEED intensity evaluation /4.3/, resulting in an on-top adsorption of the CO molecule with a tilted molecular axis, was in contradiction to the result of other methods such as photoemission experiments claiming a vertically bound molecule /4.4/. Meanwhile careful LEED reinvestigations avoiding surface damage by the primary beam did also result in this structure /2.9,4.5-8/. The sensitivity with respect to a primary beam current of about 3 µA is demonstrated in Fig.4.6 /2.9/ for the 1/2 1/2 beam measured with the fast TV method taking 16 s for a spectrum of 800 points. The upper curve is taken immediately after the c(2×2) superstructure has developed, the following ones 100 s, 5 min, and 15 min later, respectively. It appears that already after 100 s considerable changes occur. The curves recorded even later are certainly not suited to be used in a theory-experiment fit. The integer beam spectra are also affected, but less severely as demonstrated for the 10 beam in Fig.4.7 /2.9/.

5. New Possibilities Using Modern Intensity Measurement Methods

The availability of high-speed intensity measurement methods with automatic background subtraction and immediate feed back not only helps to improve measurements but also opens new ways of data collection. These, however, can only be sketched until future development will show and utilize all their advantages.

5.1 Integral Intensities of Rapidly Varying Surface Systems

The development of a surface structure during an adsorption process could not be investigated from diffraction intensities so far because of the speed of adsorption. The same holds for the interesting field of phase transitions of clean surfaces, which can be semiconductors (e.g., Si, GaAs) as well as metals (e.g., Mo, W, Ir, Pt, Au). From the recording of complete intensity spectra of such ordering processes information about the structure of intermediate states may be deduced. However, in order to guarantee that a spectrum refers to one definite state of the surface for each energy range, the time necessary for the measurement must be considerably smaller than the time constant of the process under development. This can

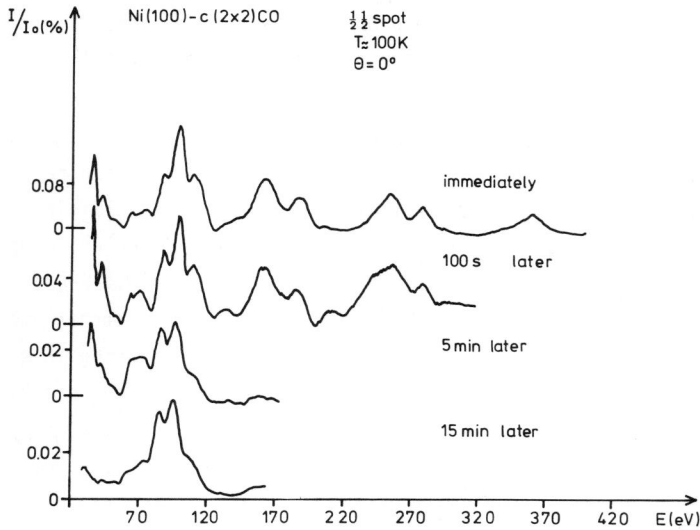

Fig.4.6

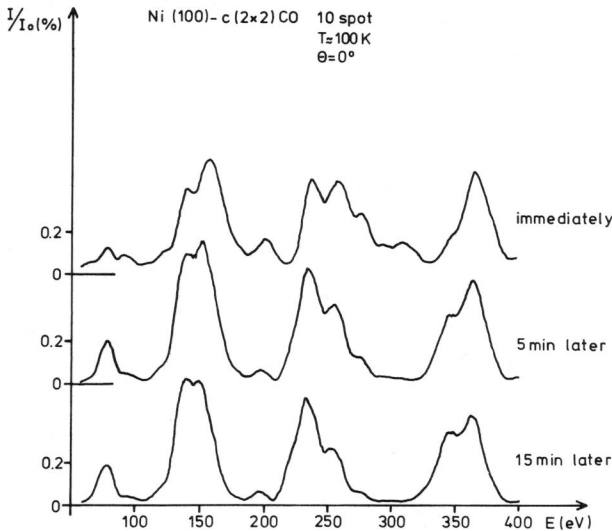

Fig.4.7

Fig.4.6. Influence of the primary beam demonstrated for the CO adsorption on Ni(100) /2.9/. The upper curve was taken within 16 s immediately after the c(2×2) structure had been fully developed. The following curves were measured with the same speed, however, with a time delay during which the primary beam was kept impinging onto the sample. Considerable structural changes occur already after 100 s

Fig.4.7. Same as Fig.4.6, but for the 10 beam

Fig.5.1. Behavior of the 10 spectrum of W(100) near room temperature and for normal incidence during hydrogen adsorption at 2.5×10^{-7} Pa. The LEED patterns were taken at 44 eV and each spectrum measured within about 13 s. The curve on top corresponds to the clean surface. Modifications of the spectrum are detectable already in the very beginning of the adsorption process which ends in a 1×1 adsorption structure /3.15/

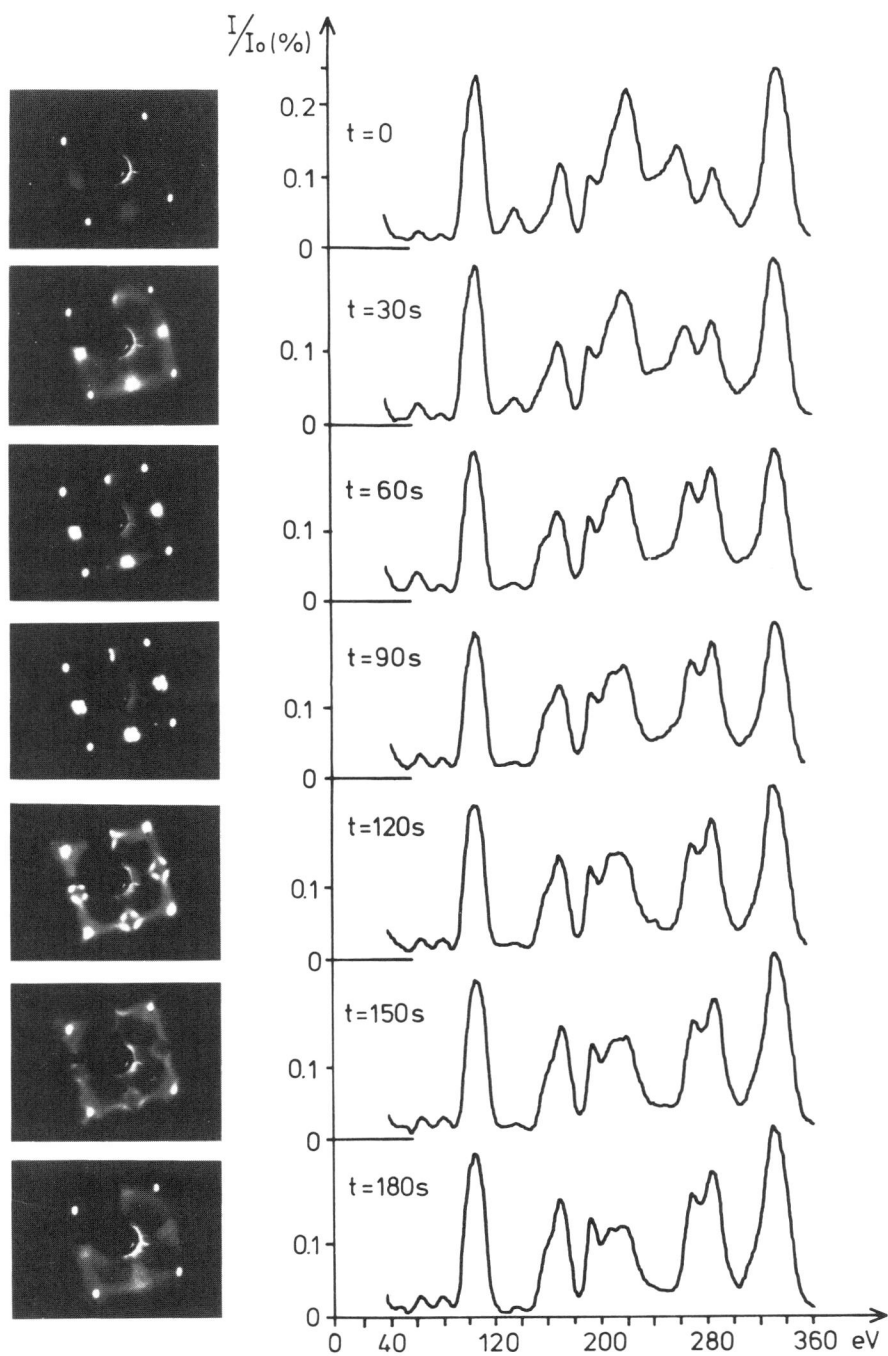

Fig.5.1. Figure caption see opposite page

be achieved by the very fast TV method described. In Fig.5.1 the study of the adsorption process of hydrogen on W(100) is demonstrated for room temperature and a
hydrogen pressure of 2.5×10^{-7} Pa /2.22/. Together with the LEED patterns the spectra of the 10 beams for normal incidence of the primary beam are given every 30 s
beginning with the clean surface (t = 0). The measurement of each spectrum took
about 13 s. The c(2×2) adsorption superstructure is fully developed after 1 min
followed by a continuous splitting of the half order beams which finally disappear
to give a (1×1) adsorption structure. It appears that considerable changes already
occur after 30 s of adsorption corresponding to about a 0.1 coverage as can be seen
from the peak group at 280 eV. At higher coverages the region about 200 eV is affected also. However, at the moment theory allows no conclusion from the data because dynamic intensities of only partly ordered systems are difficult to interpret.

5.2 Spot Profiles of Rapidly Varying Surface Systems

As the profiles of diffraction spots are based more on kinematic grounds than under
integral spot intensities, it is promising to observe profiles of spots which appear
or disappear during adsorption or a surface phase transition. This can be easily
performed with the TV method too, because the extraction of integral intensities is
based on the measurement of spot profiles. An example is demonstrated in Fig.5.2
(partly taken from /2.22/) for W(100) which shows a (1×1)→c(2×2) transition with
decreasing temperature. On the left-hand side the pattern of the initial and final
structure are displayed. The development of the half order spot profiles is given
on the right-hand side for various temperatures. The upper curve shows the background distribution in the spot region and demonstrates once more the importance
of background subtraction in order to yield reliable data. The profiles correspond
to one TV line through the spot maximum which can be taken within 1/50 s, so the
measurement matches the change of temperature. In the example of Fig.5.2, the decrease from 773 K down to 159 K took place within only a few seconds. Especially
for the example of the described W(100) phase transition, this speed is of considerable importance because it safely avoids hydrogen adsorption from the residual
gas, which for a long time had been suspected to cause the c(2×2) superstructure.
Because of the extreme speed, the temperature values where the measurements have
actually been performed are accidental to some extent. It appears that the 1/2 1/2
spot begins to appear at about 370 K and grows during decreasing temperature with
narrowing half width. The half width can be used to extract the diameter of the developing c(2×2) domains for each temperature. As demonstrated in /2.21,22/ it turns
out that the half width of the 1/2 1/2 spot does not change any more below 200 K,

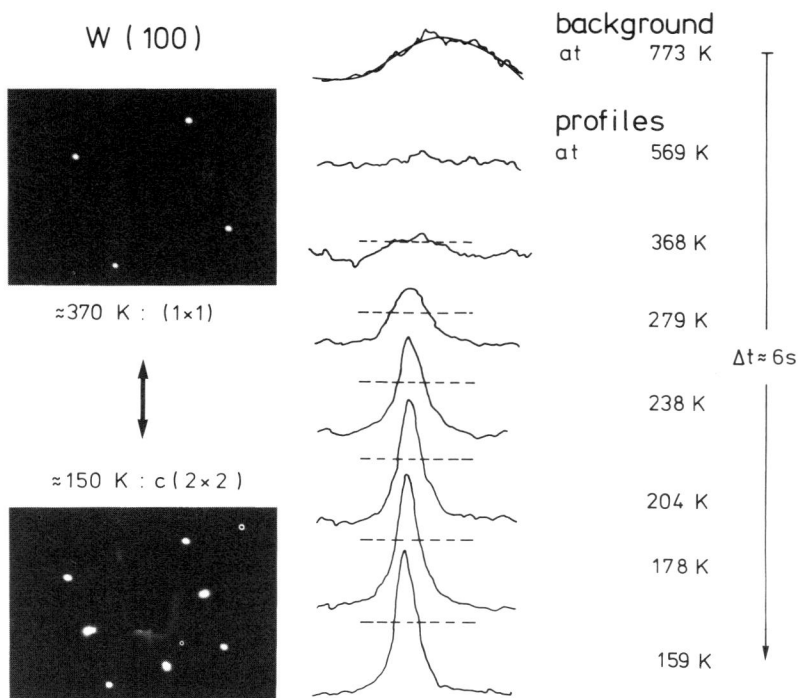

Fig.5.2. Development of the half order spot profile at 43 eV during the temperature phase transition of W(100) 1×1→c(2×2). The measurements with the TV method could be performed very rapidly as to exclude possible hydrogen adsorption

corresponding to a domain diameter of about 150 Å. However, it must be assumed that this limitation is due to the coherence width of the incident beam rather than to a limitation of the domain growth. This interpretation is supported by the behavior of the work function which is still changing at this point /2.21/.

Of course there is more information hidden in spot profiles than extracted by the evaluation of spot width only, e.g., the extent and kind of defects and disorder. A more comprehensive representation is given in /2.30/. Especially the dynamic observation of spot splitting as, e.g., in the case of hydrogen adsorption on W(100), described in the preceeding section could help to interpret the process of adsorption.

5.3 Extension of Intensity Measurements to Varying Temperature

In most cases intensity measurements are performed for fixed temperatures. The influence of the temperature is considered in the dynamic calculation by complex phase

shifts which simulate a Debye-Waller factor for a given temperature and Debye temperature. However, the latter is known only for the bulk and its modification at the
surface due to the truncation of part of the bonds is only roughly known. So, for
the careful test of surface models, the surface Debye temperature has to be varied
until a best fit is achieved. However, the experience of many works of this kind is
that this procedure is very insensitive and no definite conclusion for the surface
Debye temperature can be drawn.

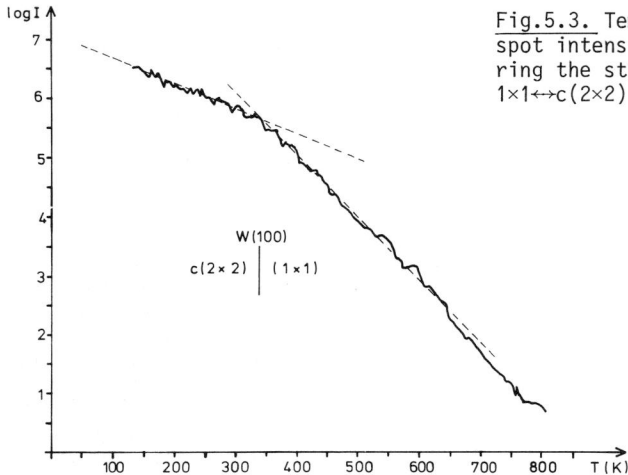

Fig.5.3. Temperature dependence of 10 spot intensity of W(100) at 142 eV during the structural phase transition 1×1↔c(2×2) /2.22/

If, however, the temperature dependence of intensities could be measured separately, much more information for the determination of the surface Debye temperature would be available. This implies an immense amount of data collection work because intensity spectra have to be measured for many temperatures. Alternatively intensities have to be collected as a function of temperature for many different energies, which is possible with a very fast data collection system. This is demonstrated
for one energy of the 10 beam of W(100) in Fig.5.3. The change of slope corresponding
to a change of the surface Debye temperature is apparent in the region where the
phase transition takes place. By assuming kinematic diffraction only, a Debye temperature can be extracted from these data for each energy. As a kinematic interpretation is not at all true for LEED and because of the mixing of surface layer and
bulk layer diffraction with the penetration of the incident electrons, the resulting
"effective" Debye temperature varies strongly with energy. This is demonstrated in
Fig.5.4 for the two phases of the clean W(100) surface. It is obvious that it is
most dangerous to take the result for Θ_{eff} from a measurement at a single energy,
e.g., an assumed Bragg peak energy as sometimes done to determine the Debye tempe-

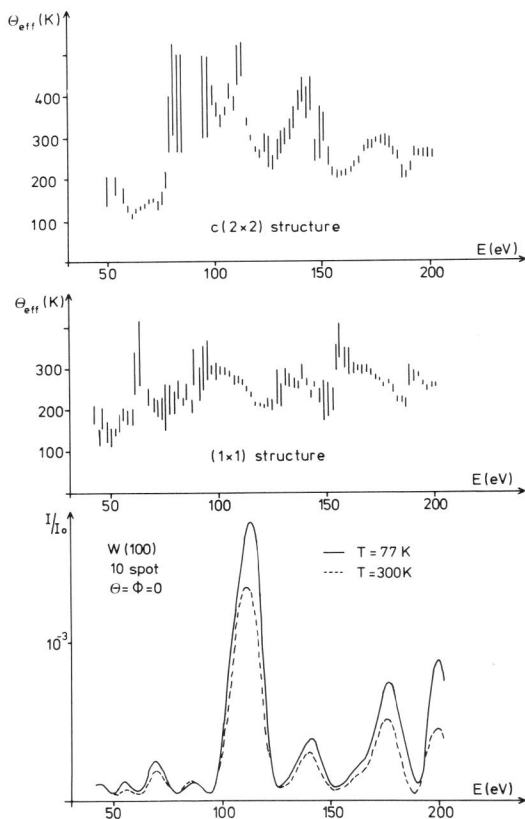

Fig.5.4. Effective surface Debye temperature resulting from kinematic evaluation of intensity temperature data of the 10 spot for the c(2×2) and 1×1 phases of W(100). The data are evaluated from sets of measurement as in Fig.5.3. The lower curves demonstrate also that a simple Debye-Waller factor is not suited to correct intensity spectra for temperature effects

rature. Moreover, the separation of the physical surface Debye temperature should be performed by varying this parameter in dynamic calculations and fitting the results to the temperature spectra shown in Fig.5.4. This has been done in /2.21/, resulting in values of 400 K and 210 K for the surface Debye temperatures of the c(2×2) and (1×1) structures of W(100), respectively. However, these values were given with large error bars of ± 100 K and ± 40 K, respectively, which are due to the very fast measurement speed used to avoid disturbing hydrogen adsorption.

5.4 Extension of Intensity Measurements to the Medium-Energy Range

The electron diffraction intensities become gradually more kinematic with increasing energy (e.g., /5.1/), and therefore, simpler theoretical methods for data interpretation might be expected. However, the range of energy in which data acquisition has to be collected then grows to several hundreds of eV, increasing correspondingly the

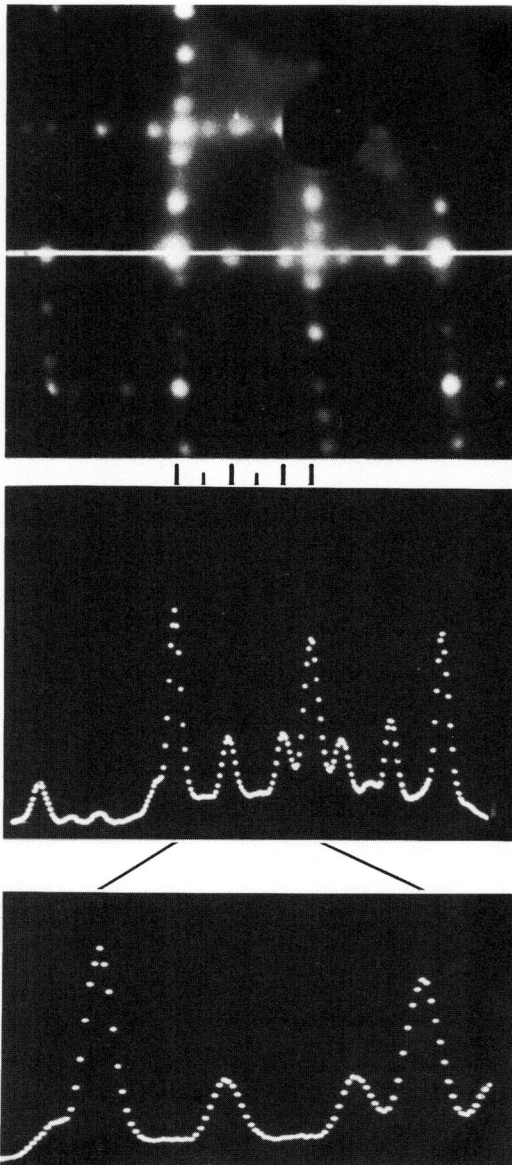

Ir (100) 5×1   T = 100 K

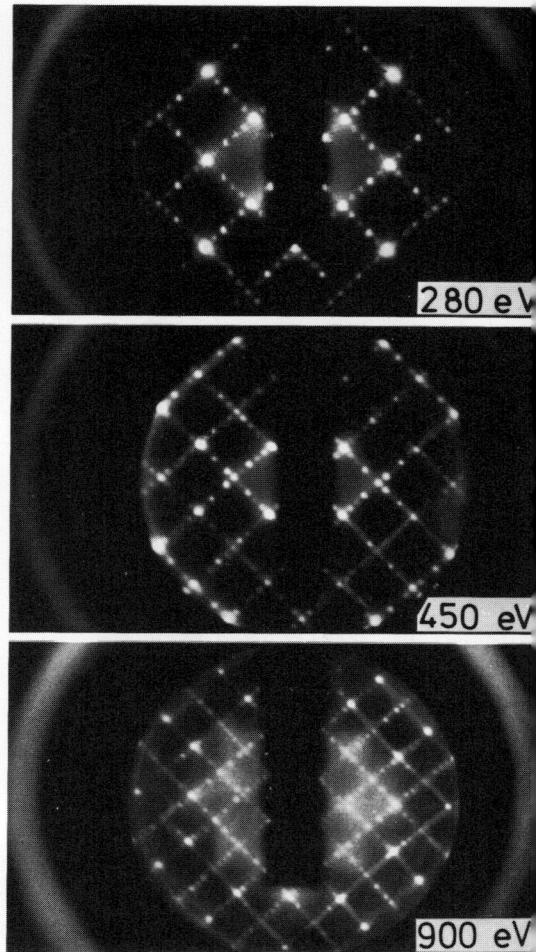

280 eV

450 eV

900 eV

Fig.5.6

Fig.5.5

Fig.5.5. Lateral resolution of the TV system on the monitor. In the middle frame the intensity distribution along the line shaped electronic window within the up-per frame (pattern of Ir(100) 5×1 at 365 eV) is shown, at the bottom frame with an expanded scale. The distance of neighboring points corresponds to a diffraction angle resolution of 0.3°

Fig.5.6. LEED pattern of Ir(100) 5×1 at T ~ 100 K for 280, 450 and 900 eV. It ap-pears that even at the highest energy the surface sensitivity is not lost as broken order spots show up and can be resolved

time of measurement needed. Moreover, at high energies a lot of diffraction beams
show up and have to be taken into account. Last, but not least, the measurement itself becomes more complicated as the background level increases considerably and
the spots crowd together, at least in the case of complicated superstructures. However, the fast TV method for intensity measurements handles all these experimental
disadvantages. The increase of data collection because of the increased number of
beams is no problem because of the speed of measurement and for the possibility to
use a video tape recorder. The background level increase can at least partially be
taken care of by the automatic background subtraction, but can and should additionally be reduced by lowering the temperature of the sample. The problem of high-energy crowding of diffraction spots is less important in view of the spatial resolution
of the video system. This is of the order of 0.3° as demonstrated in Fig.5.5 for the
1×5 superstructure of clean Ir(100). The electronic window is chosen to degenerate
to a linear cut (upper frame) along which the intensity distribution is given in the
middle and part of it below at an extended scale. The value of 0.3° corresponds to
the angular separation of the digitizing data points. The experience is that intensity spectra can be measured as long as the diffraction spots appear to be resolved
on the luminescent screen. In Fig.5.6 the pattern of Ir(100) 5×1 is once more shown
for low, medium and higher energy for demonstration. Figure 5.7 shows the result of
the measurement for the 0 3/5 beam up to an energy of 550 eV. It should be pointed
out that there is much structure in the medium-energy range, more than in the low-
energy range below 200 eV. This looks encouraging for the extraction of information
with respect to surface structure.

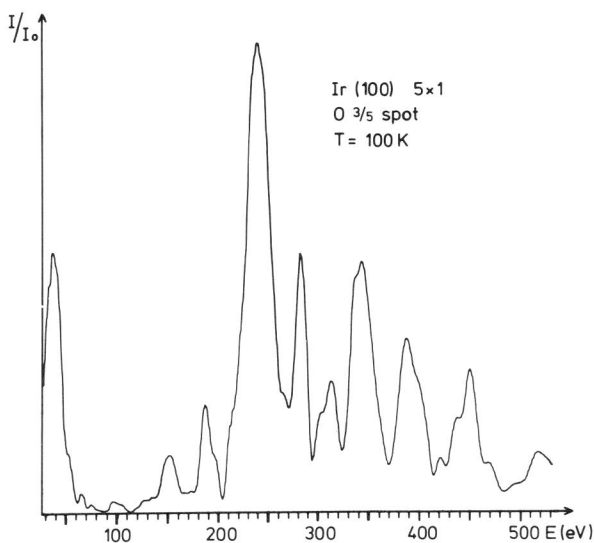

Fig.5.7. Intensity spectrum of
the 0 3/5 spot of Ir(100) 5 × 1
at T ∼ 100 K and normal incidence
up to the medium energy range

6. Summary and Outlook

The development of successful schemes for dynamic LEED intensity calculations has stimulated a variety of activities to improve the reliability of experimental LEED data. This appeared to be necessary in order to extract better structural information based on precise data. While the comparison of calculated spectra and experimental results use to be a matter of visual inspection, it is currently being done on a quantitative scale represented by reliability factors which consider essentially the peak structure of intensity spectra. In this way the theory-experiment comparison is now extended also to minor features of the spectra which are not necessarily identical in the results of different laboratories. In some cases data were reported that differed considerably under visual inspection, but were nevertheless believed to result from the same surface conditions. In nearly all cases the discrepancies appear to be caused by a set of only a few experimental uncertainties such as surface misalignment, residual gas adsorption, adsorbate decomposition or desorption, and missing or incomplete background subtraction.

The aim of all instrumental activities in LEED was to overcome these influences. They resulted in the development of new measurement techniques with different advantages over the classical methods, which make a Faraday cup or spot photometer track a diffraction spot moving with varying energy or angle of incidence. An increased speed of measurement promises to avoid time effects arising from residual gas adsorption or adsorbate decomposition/desorption by the primary beam. An increased ease in the measurement procedure together with high speed reduces control measurements to the level of routine which is desirable to avoid misalignment. The possibility to overcome systematic errors of the measurement caused by incomplete background subtraction depends strongly on the measurement procedure itself.

The new developments can be grouped in technical improvements of the classical methods using an automatic spot tracking procedure and in techniques based on different concepts such as the photographic and the TV computer methods. At the moment it appears that the TV computer methods make the most rapid on-line measurements possible and allow for careful on-line background subtraction and normalization with respect to the primary beam. Sample misalignment off normal incidence can be reduced below practical relevance by successive readjustments until the spectra of equivalent beams differ only to a negligible degree, a procedure which is practicable because of the speed and ease of the technique. High-precision measurements can be taken in the order of seconds leaving residual gas adsorption or surface distortion by the primary beam negligible.

Recent experimental progress provides a safe basis for an efficient theory-experiment fit. Theoretical progress is, however, highly desirable to support the evaluation of data which now can be taken using the new techniques. The good spatial resolution makes intensity spectra of complicated LEED patterns with a large number of spots available, as well as intensities and spot profiles of time-varying systems such as structural phase transitions and surface reactions. However, it is the authors' impression that a development of LEED theory to more and more complex and multiparametrized schemes of calculation should not be expected to be able to extract surface information alone. It seems also promising to look for special LEED data which can be evaluated in an easier way as is already done on a minor scale by low-temperature measurements to avoid the influence of thermal vibrations. More effectively, this can be extended to data from the medium-energy range which can be easily recorded over some hundreds of eV and which are still surface sensitive. The evaluation of these data is facilitated by kinematic approximation of the layer diffraction with eventually some dynamic corrections. Though early efforts in this direction were not fully successful, it seems promising to do some further development in this field. In other cases as in surface-ordering processes, e.g., during adsorption or surface phase transitions, the evaluation of spot profiles should be more favorable than that of absolute intensities. They can be measured in rapid sequences during the dynamic change of the surface and are more easily interpreted than absolute intensities because their shape is dominated by kinematic diffraction. Perhaps, with increasing spread of more sophisticated experiments with extended possibilities, other methods for the evaluation of LEED data will be developed also.

Acknowledgements: We are indebted to those members of our laboratory who have contributed to this paper by their experience and their results.

References

1.1   H.H. Brongersma: J. Vac. Sci. Technol. *11*, 231 (1974)
1.2   F.W. Savis, J.F. van der Veen: "Analysis of Surface Structure and Composition by Ion-Scattering Spectroscopy", in *Proc. 7th Intern. Vac. Congr. & 3rd Intern. Conf. Solid Surfaces* (Vienna 1977) p. 2503
1.3   W. Heiland, E. Taglauer: Surf. Sci. *69*, 96 (1977)
1.4   T. Engel, K.H. Rieder: this volume
1.5   H. Wilsch: "Atomic and Molecular Scattering from Surfaces-Elastic Scattering", ed. by E. Kay, P.S. Bagus, in *Topics in Surface Chemistry* (Plenum, New York 1978) p. 135
1.6   N.V. Smith: "Angular Dependent Photoemission", in *Photoemission in Solids I*, ed. by M. Cardona, L. Ley, Topics in Applied Physics Vol. 26 (Springer, Berlin, Heidelberg, New York 1978) p. 237
1.7   B. Feuerbacher, B. Fitton, R.F. Willis (eds.): *Photoemission and the Electronic Properties of Surfaces* (John Wiley & Sons, New York 1978)
1.8   D.P. Woodruff: Surf. Sci. *53*, 538 (1975)

1.9   J.W. Gadzuk: Surf. Sci. *60*, 76 (1976)
1.10   S.P. Weeks, A. Liebsch: Surf. Sci. *62*, 197 (1977)
1.11   D. Aberdam, R. Baudoing, E. Blanc, C. Gaubert: Surf. Sci. *65*, 77 (1977);
       Surf. Sci. *71*, 279 (1978)
1.12   J.D. Place, M. Prutton: Surf. Sci. *82*, 315 (1979)
1.13   T. Matsudaira, M. Onchi: J. Phys. C *12*, 3381 (1979)
1.14   C.J. Davison, L.H. Germer: Phys. Rev. *30*, 305 (1927)
1.15   J.J. Lander: Progr. Solid State Chem. *2*, 26 (1965)
1.16   E. Bauer: "Techniques for the Direct Observation of Structure and Imperfec-
       tions", in *Techniques of Materials Research*, ed. by R.F. Bunshaw (John Wiley
       & Sons, New York 1969) p. 559
1.17   J.W. May: Ad. Catal. *21*, 152 (1970)
1.18   P.J. Estrup, E.G. McRae: Surf. Sci. *25*, 1 (1971)
1.19   G.A. Somorjai, H.H. Farrell: Ad. Chem. Phys. *20*, 215 (1972)
1.20   M. Láznička (ed.): *LEED-Surface Structures of Solids*, (Union of Czechoslovak
       Mathematicians and Physicists, Prague 1972)
1.21   M.B. Webb and M.G. Lagally: "Elastic Scattering of Low-Energy Electrons from
       Surfaces", in *Solid State Physics 28*, ed. by H. Ehrenreich et al. (Academic
       Press, New York, London 1973) p. 301
1.22   C.B. Duke: "Determination of the Structure and Properties of Solid Surfaces
       by Electron Diffraction and Emission", in *Aspects of the Study of Surfaces*,
       ed. by I. Prigogines, S.A. Rice; Advances in Chemical Physics XXVII (John
       Wiley & Sons, New York 1974) p. 1
1.23   G. Ertl, J. Küppers: "Low Energy Electrons and Surface Chemistry" (Verlag
       Chemie, Weinheim 1974)
1.24   E.G. Bauer: "Low Energy Electron Diffraction (LEED) and Auger Methods", in
       Interactions on Metal Surfaces, ed. by R. Gomer, Topics Appl. Phys. *4*
       (Springer, Berlin, Heidelberg, New York 1975) p. 225
1.25   J.A. Strozier, JR., D.W. Jepsen, F. Jona: "Surface Crystallography", in
       Surface Physics of Materials I, ed. by J.M. Blakely (Academic Press, New York,
       London 1975) p. 1
1.26   P.M. Marcus: "Surface Structure by Analysis of LEED Intensity Measurements",
       in *Characterization of Metal and Polymer Surfaces I*, ed. by L.H. Lee (Aca-
       demic Press, New York, London 1977) p. 271
1.27   P.J. Estrup: "LEED Studies of Surface Layers", in *Characterization of Metal
       and Polymer Surfaces I*, ed. by L.H. Lee (Academic Press, New York, London
       1977) p. 187
1.28   F. Jona: Surf. Sci. *68*, 204 (1977)
1.29   S.Y. Tong: "A Review of Surface Crystallography by Low-Energy Electron Dif-
       fraction", in *Electron Diffraction 1927-1977*, ed. by P.J. Dobson, J.B. Pendry,
       C.J. Humphreys (The Institute of Physics, Bristol, London 1978) p. 270
1.30   F. Jona: J. Phys. C: Solid State Phys. *11*, 4271 (1978)
1.31   J.B. Pendry: "Low Energy Electron Diffraction" (Academic Press, New York,
       London 1974)
1.32   M.A. Van Hove, S.Y. Tong: "Surface Crystallography by LEED", Springer Ser.
       Chem. Phys. *2*, (Springer, Berlin, Heidelberg, New York 1979)
1.33   M.A. Van Hove: "Surface Crystallography and Bonding", in *The Nature of the
       Surface Chemical Bond*, ed. by T.N. Rhodin, G. Ertl (North Holland, Amsterdam,
       New York, Oxford 1979) p. 275
1.34   G.A. Somorjai, M.A. Van Hove: "Adsorbed Monolayers on Solid Surfaces",
       Structure and Bonding *38* (Springer, Berlin, Heidelberg, New York 1979)

2.1   Groupe d'étude des surfaces (D. Aberdam, R. Baudoing, C. Gaubert, J. Gauthier,
       V. Hoffstein): Surf. Sci. *48*, 496 (1975)
2.2   J. Gauthier, D. Aberdam, R. Baudoing: Surf. Sci. *78*, 339 (1978)
2.3   J.E. Demuth, D.W. Jepsen, P.M. Marcus: Solid State Commun. *13*, 1311 (1973)
2.4   P.M. Marcus, J.E. Demuth, D.W. Jepsen: Surf. Sci. *53*, 501 (1975)
2.5   M.A. Van Hove, S.Y. Tong: Phys. Rev. Lett. *35*, 1092 (1975)
2.6   M.A. Van Hove, S.Y. Tong, M.H. Elconin: Surf. Sci. *64*, 85 (1977)

2.7   K.O. Legg, F. Jona, D.W. Jepsen, P.M. Marcus: Phys. Rev. B *16*, 5271 (1977)
2.8   E. Zanazzi, F. Jona: Surf. Sci. *62*, 61 (1977)
2.9   K. Heinz, E. Lang, K. Müller: Surf. Sci. *87*, 595 (1979)
2.10 M. Passler, A. Ignatiev, F. Jona, D.W. Jepsen P.M. Marcus: Phys. Rev. Lett. *43*, 360 (1979)
2.11 S. Andersson, J.B. Pendry: Phys. Rev. Lett. *43*, 363 (1979); J. Phys. C: Solid State Phys. *13*, 3547 (1980)
2.12 S.Y. Tong, A. Moldonando, C.H. Li, M.A. Van Hove: Surf. Sci. *94*, 73 (1980)
2.13 J.B. Pendry: J. Phys. C: Solid State Phys. *13*, 937 (1980)
2.14 P. Marcus, F. Jona (ed.): *Proc. Conf. on Determination of Surface Structure by LEED* (Yorktown Heights 1980)
2.15 J.R. Noonan, H.L. Davis: J. Vac. Sci. Technol. *17*, 194 (1980)
2.16 M.N. Read, G.J. Russel: Surf. Sci. *88*, 95 (1979)
2.17 M. Kalisvaart, M.R. O'Neill, T.W. Riddle, F.B. Dunning, G.K. Walters: Phys. Rev. *17*, 1570 (1978)
2.18 M.K. Debe, D.A. King, F.S. Marsh: Surf. Sci. *68*, 437 (1977)
2.19 B.W. Lee, A. Ignatiev, S.Y. Tong, M.A. Van Hove: J. Vac. Sci. Technol. *14*, 291 (1977)
2.20 P.S.P. Wei: J. Chem. Phys. *53*, 2939 (1970)
2.21 P. Heilmann, K. Heinz, K. Müller: Surf. Sci. *89*, 84 (1979)
2.22 P. Heilmann: Dissertation (Erlangen 1979)
2.23 M.A. Van Hove, S.Y. Tong: Surf. Sci. *54*, 91 (1976)
2.24 J. Kirschner, R. Feder: Surf. Sci. *79*, 176 (1979)
2.25 F.S. Marsh, M.K. Debe, D.A. King: J. Phys. C: Solid State Phys. *13*, 2799 (1980)
2.26 E. Lang, K. Heinz, to be published
2.27 G.E. Laramore, J.E. Houston, R.L. Park: J. Vac. Sci. Technol. *10*, 196 (1973)
2.28 J.E. Houston, G.E. Laramore, R.L. Park: Surf. Sci. *34*, 477 (1973)
2.29 H. Wagner: "Physical and Chemical Properties of Stepped Surfaces", in *Solid Surface Physics*, Springer Tracts in Modern Physics *85*, ed. by G. Höhler, E.A. Niekisch (Springer, Berlin, Heidelberg, New York 1979) p. 151
2.30 M. Henzler: "Electron Diffraction and Surface Defect Structure", in *Electron Spectroscopy for Surface Analysis*, Topics in Current Physics *4*, ed. by H. Ibach (Springer, Berlin, Heidelberg, New York 1977) p. 116
2.31 G.E. Rhead: Surf. Sci. *68*, 20 (1977)
2.32 J. Jagodzinski: "Diffraction in Surfaces and Interfaces", in *Festkörperprobleme*, Advances in Solid State Physics, Vol. XVIII (Vieweg, Braunschweig 1978) p. 129
2.33 J. Jagodzinski, W. Moritz, D. Wolf: Surf. Sci. *77*, 233 (1978)
2.34 W. Moritz, H. Jagodzinski, D. Wolf: Surf. Sci. *77*, 249 (1978)
2.35 D. Wolf, H. Jagodzinski, W. Moritz: Surf. Sci. *77*, 265 (1978)
2.36 D. Wolf, H. Jagodzinski, W. Moritz: Surf. Sci. *77*, 283 (1978)
2.37 J.E. Demuth, S.Y. Tong, T.N. Rhodin: J. Vac. Sci. Technol. *9*, 639 (1971)
2.38 J.E. Demuth, T.N. Rhodin: Surf. Sci. *42*, 261 (1974)
2.39 K. Müller, E. Lang, L. Hammer, W. Grimm, P. Heilmann, K. Heinz: *Proc. Conf. on Determination of Surface Structure by LEED* (Yorktown Heights 1980)
2.40 J.R. Noonan, H.L. Davis: *Proc. Conf. on Determination of Surface Structure by LEED* (Yorktown Heights 1980)
2.41 H. Eckart: Diplomarbeit (Erlangen 1976)
2.42 T.M. Madey, J.T. Yates, Jr.: J. Vac. Sci. Technol. *8*, 525 (1971)
2.43 C.J. Blunette, L.W. Swanson: J. Appl. Phys. *39*, 2749 (1968)
2.44 D. Menzel: "Desorption Phenomena", in *Interactions on Metal Surface Topics*, Applied Physics *4*, ed. by R. Gomer (Springer, Berlin, Heidelberg, New York 1975) p. 101
2.45 M. J. Drinkwine, D. Lichtman: Progr. Surf. Sci. *8*, 123 (1977)
2.46 M.L. Shek, S.P. Withrow, W.H. Weinberg: Surf. Sci. *72*, 678 (1978)
2.47 J.C. Tracy: J. Chem. Phys. *56*, 2736 (1972)
2.48 L.L. Kesmodel, P.C. Stair, R.C. Baetzold, G.A. Somorjai: Phys. Rev. Lett. *36*, 1316 (1976)

52

2.49 H.E. Farnsworth: Phys. Rev. *34*, 679 (1929)
2.50 R.L. Park, H.E. Farnsworth: Rev. Sci. Instr. *35*, 1592 (1964)
2.51 Ch. Barner, K. Müller, G.G. Waldecker: Verhandl. DPG *3*, 427 (1973)
2.52 G.G. Waldecker: Thesis (Erlangen 1976)
2.53 H.J. Bechthold, E. Kreutz, E. Rickus, N. Sotnik: Verhandl. DPG (VI) *13*, 585 (1978)
2.54 W. Moritz, D. Wolf: Surf. Sci. *88*, L29 (1979)
2.55 D. Wolf: private communication
2.56 W. Ehrenberg: Phil. Mag. *18*, 878 (1934)
2.57 K.O. Legg, M. Prutton, C. Kinniburgh: J. Phys. C: Solid State Phys. *7*, 4236 (1974)
2.58 R.E. McCarry: J. Appl. Phys. *37*, 473 (1965)
2.59 T.E. Felter, P.J. Estrup: Rev. Sci. Instrum. *47*, 158 (1966)
2.60 T. Kanaji, H. Nakatsuka, T. Urano, Y. Taki: Surf. Sci. *86*, 587 (1979)
2.61 W. Berndt: Verhandl. DPG (VI) *15*, 718 (1980)
2.62 W. Berndt: private communication
2.63 J.M. Baker, J.M. Blakely: Surf. Sci. *32*, 45 (1972)

3.1   P.C. Stair, T.J. Kaminska, L.L. Kesmodel, G.A. Somorjai: Phys. Rev. B *11*, 623 (1975)
3.2   L.L. Kesmodel, L.H. Dubois, G.A. Somorjai: J. Chem. Phys. *70*, 2180 (1979)
3.3   M.A. Van Hove, R.J. Koestner, P.C. Stair, J.P. Biberian, L.L. Kesmodel, I. Bartros, G.A. Somorjai: Surf. Sci. (in press)
3.4   K. Ueda, F. Forstmann: *Proc. 7th Intern. Vac. Congr. & 3rd Intern. Conf. Solid Surfaces* (Vienna 1977) p. 2423
3.5   D.C. Frost, K.A.R. Mitchell, F.R. Shepherd, P.R. Watson: J. Vac. Sci. Technol. *13*, 1196 (1976)
3.6   T.N. Tommet, G.B. Olszewski, P.A. Chadwick, S.L. Bernasek: Rev. Sci. Instrum. *50*, 147 (1979)
3.7   K.A.R. Mitchell, F.R. Shepherd, P.R. Watson, D.C. Frost: Surf. Sci. *64*, 737 (1977)
3.8   F.R. Shepherd, P.R. Watson, D.C. Frost, K.A.R. Mitchell: J. Phys. C: Solid State Phys. *11*, 4591 (1978)
3.9   D.C. Frost, S. Hengrasmee, K.A.R. Mitchell, F.R. Shepherd, P.R. Watson: Surf. Sci. *76*, L585 (1978)
3.10 D.C. Frost, K.A.R. Mitchell, W.T. Moore, R.W. Streater, P.R. Watson: *Proc. 7th Intern. Vac. Congr. & 3rd Intern. Conf. Solid Surfaces* (Vienna 1977) p. 2403
3.11 W.T. Moore, P.R. Watson, D.C. Frost, K.A.R. Mitchell: J. Phys. C: Solid State Phys. *12*, L887 (1979)
3.12 S. Hengrasmee, P.R. Watson, D.C. Frost, K.A.R. Mitchell: Surf. Sci. *87*, L249 (1979)
3.13 S. Hengrasmee, P.R. Watson, D.C. Frost, K.A.R. Mitchell: Surf. Sci. *92*, 71 (1980)
3.14 P. Heilmann, E. Lang, K. Heinz, K. Müller: Appl. Phys. *9*, 247 (1976)
3.15 E. Lang, P. Heilmann, G. Hanke, K. Heinz, K. Müller: Appl. Phys. *19*, 287 (1979)
3.16 J.E. Demuth, T.N. Rhodin: Surf. Sci. *42*, 261 (1974)
3.17 G. Hanke, E. Lang, K. Heinz, K. Müller: Surf. Sci. *91*, 193 (1979)
3.18 F. Jona: private communication
3.19 H. Leonhard, A. Gutmann, K. Heyek: J. Phys. E: Sci. Instrum. *13*, 288 (1980)
3.20 D.G. Welkie, M.G. Lagally: Appl. Surf. Sci. *3*, 272 (1979)
3.21 S.P. Weeks, J.E. Rowe, S.B. Christman, E.E. Chaban: Rev. Sci. Instrum. *50*, 1249 (1979)
3.22 M.D. Chinn, S.C. Fain, Jr.: J. Vac. Sci. Technol. *14*, 314 (1976)
3.23 D.G. Welkie, M.G. Lagally: J. Vac. Sci. Technol. *16*, 784 (1979)
3.24 S.P. Weeks, J.E. Rowe: J. Vac. Sci. Technol. *16*, 470 (1979)
3.25 P. Heilmann, E. Lang, K. Heinz, K. Müller: *Proc. Conf. on Determination of Surface Structure by LEED* (Yorktown Heights 1980)

4.1 M.K. Debe, D.A. King: Surf. Sci. *81*, 193 (1979)
4.2 J.E. Demuth, T.N. Rhodin: Surf. Sci. *45*, 249 (1974)
4.3 S. Andersson, J.B. Pendry: Surf. Sci. *71*, 75 (1978)
4.4 C.L. Allyn, T. Gustafsson, E. Plummer: Solid State Commun. *28*, 85 (1978)
4.5 M. Paessler, A. Ignatiev, F. Jona, D.W. Jepsen, P.M. Marcus: Phys. Rev. Lett. *43*, 360 (1979)
4.6 S. Andersson, J.B. Pendry: Phys. Rev. Lett. *43*, 363 (1979)
4.7 S. Andersson, J.B. Pendry: J. Phys. C: Solid State Phys. *13*, 3547 (1980)
4.8 S.Y. Tong, A. Maldonado, C.H. Li, M.A. Van Hove: Surf. Sci. *94*, 73 (1980)

5.1 D. Aberdam: "Electron Diffraction in the Medium-Energy Range", in *Electron Diffraction 1927–1977*, ed. by P.J. Dobson, J.B. Pendry, C.J. Humphreys (The Institute of Physics, Bristol, London 1978) p. 239

Structural Studies of Surfaces with Atomic and Molecular Beam Diffraction

T. Engel and K. H. Rieder

1. Introduction

The geometrical arrangement of the atoms in the topmost layer of a solid may differ from that in the bulk due to the symmetry-breaking nature of the surface. This has important consequences for the reactivity and electronic properties of the solid in question; for this reason, structural studies on solid surfaces are a major research area in surface science. In the past decade, a substantial number of investigations using primarily LEED /1.1/ has significantly increased our understanding of the atomic arrangement of clean and adsorbate-covered surfaces. However, because of the relatively high energy of the electrons used in LEED (20-200 eV), the electron scattering occurs almost exclusively from the ion cores, whereas the scattering from the valence-electron distribution is weak. Since a better understanding of the reactivity and electronic properties of surfaces requires a knowledge of both the atomic positions and the electron distribution, the knowledge gained with LEED must be supplemented by that obtained with another technique more sensitive to the electron-charge distribution. This is particularly important if the chemisorption bond at surfaces is to be better understood.

Atomic and molecular-beam diffraction from surfaces is a technique which can fulfill this criterion. Although first experiments by ESTERMANN and STERN /1.2/ successfully showed in 1928 that atom diffraction could be observed, the technique has been applied to surfaces other than LiF(100) only in the past few years. However, in this short time period a considerable advance has been made in both our understanding of the fundamental interaction between an incoming particle and the surface, and in our picture of the electron-charge distribution at surfaces. This review will discuss these developments and their relevance with regard to our understanding of surface science. Before proceeding to a description of the contents of this review, we shall emphasize those aspects of atomic and molecular-beam diffraction which make it an attractive addition to the already broad spectrum of surface-science techniques.

Since the pioneering studies of Estermann and Stern, diffraction has been observed for He, Ne, H_2, D_2, HD, H, and D from a number of surfaces. Helium is the particle most widely used in diffractive investigations because of the enhanced elastic scattering with respect to neon due to its lower mass (Chap.4) and because of its chemical inertness relative to atomic and molecular hydrogen and its iso- topes. This latter feature is particularly important when studying metal and semi- conductor surfaces. An especially attractive feature of atom diffraction is that for the energies used in diffraction studies (10-200 meV), no surface penetration occurs for reasonably closely packed surfaces. This gives the technique an unparal- leled surface sensitivity. It is especially valuable in surface studies in which a reconstruction involves more than one layer, since the structural information gained with atom diffraction on the topmost layer can be integrated into a LEED calculation which takes the structure of deeper layers into account. A second advantage of the technique is its completely nondestructive character. Due to the chemical inertness of He and the low impact energies used, the surface will not be damaged. Finally, an apparatus designed to carry out diffractive investigations can also be used to study inelastic processes such as the excitation of surface phonons /1.3/ and thermal ac- commodation /1.4/ as well as to study sticking /1.5/, thermal desorption /1.5/, and chemical reactions on surfaces /1.6/. A number of reviews of this field has appeared emphasizing both experimental and theoretical aspects /1.7-15/.

We begin our review in Chap.2 with a discussion of the particle-surface interac- tion potential. Both theoretical approaches to first-principles calculations of this potential and experimental determinations of the energy levels in the well are re- viewed. In Chap.3 we proceed to a discussion of the quantum-mechanical methods de- veloped to calculate diffraction intensities from a given surface corrugation. The main emphasis is placed on the hard-corrugated-wall model which, to date, has been successfully used in a number of structural investigations. This treatment first assumes a rigid lattice at rest, thermal vibrations and their influence on the dif- fraction intensities being introduced in Chap.4. A further step towards a realistic diffraction theory is made in Chap.5 in which an attractive potential is added to the hard repulsive wall. This leads to resonant scattering from the bound states of the potential which has been used experimentally to determine the bound states in the particle-surface potential. In Chap.6 we begin our review of diffraction experi- ments with a discussion of the experimental aspects of gas-surface scattering. Noz- zle-beam sources, detectors, and data treatment are discussed here. Chapter 7 re- views the experiments carried out on ionic crystal surfaces, and Chaps.8-10 discuss structural determinations with atom diffraction on semiconductor, metal, and adsor- bate-covered surfaces. Recent results on semiconductor and adsorbate-covered sur- faces show the depth of information which can be obtained with the technique.

2. The Particle–Surface Interaction Potential

2.1 Physical Basis

The use of neutral-particle diffraction (elastic scattering) for surface structural investigations requires an understanding of the basic physical principles governing the interaction of the incoming particles with the solid surfaces. We restrict our discussion to cases in which the particles and the solid do not interact chemically. Typical interaction energies are therefore in the range of physisorption energies, i.e., several tens of meV. One way to obtain an at least qualitatively correct picture of the particle-surface potential is to start from individual particle-atom potentials (for example, of the Lennard-Jones type) and to perform a summation over all binary interactions between the particle and the atoms in the lattice. The result of such a calculation for a monoatomic solid is shown in Fig.2.1 /2.1,1.7/. It illustrates schematically the most important features of a typical particle-surface potential. Consider a particle with coordinate $\underline{r}$ = $(\underline{R},z)$ outside a single crystal. The surface normal is chosen in the direction of z, so that $\underline{R}$ = (x,y) lies in the plane of the surface. At not too large distances from the surface, the particle will feel a Van der Waals' attraction due to quantum fluctuations of the charge distributions leading to fluctuating (electric) dipole-induced dipole interactions between the particle and many atoms in the crystal. Between surface atoms (A in Fig.2.1), the particle will usually have a lower potential energy than in on-top positions (B in Fig.2.1), since in the first case it feels the long-range attraction from the largest number of atoms in the solid. Closer to the surface the particle will be repelled due to short-range exchange forces arising from the overlap of the electron-density distributions, whereby the outermost valence electrons play the dominant role. The steeply rising repulsive potential will again depend on the locus of impact. For a given distance z from the topmost layer, the electron density as a function of $\underline{R}$ will have a larger value in on-top positions than between surface atoms. The periodic moldulation of the repulsive part will therefore reflect the valence-electron density at the surface which in principle contains information on both surface structure and surface bonding. Hence, for ionic crystals, where the electrons are well-localized at every ion, we expect a strong modulation of the repulsive potential parallel to the surface due to the differences of ionic radii. In contrast to that class of substances, metal surfaces will appear very smooth since the quasi-free valence electrons tend to smear out the electron distribution parallel to the surface. Strong diffraction features have indeed been observed for alkali-halide cleavage surfaces, whereas only very weak diffraction has been reported for clean low-index metal surfaces (Chaps.7,9).

The particle-surface potential function can be represented by a Fourier series,

$$V(\underline{r}) = \sum_{\underline{G}} v_{\underline{G}}(z) e^{i\underline{G}\underline{R}} \quad , \tag{2.1}$$

whereby the reciprocal lattice vectors $\underline{G}$ are defined according to (3.3,4).

The first term $v_{00}(z)$ in the series (2.1) represents a lateral average over the particle-surface interaction potential. Only discrete bound states of the particle are allowed in the attractive well, and their energies determine the angular location of resonant scattering features in diffraction experiments as discussed in Sect.5.2 (compare Fig.5.2). In Table 2.1 we list a representative cross-section of experimentally determined bound-state energy levels. As discussed also in Sect.5.2, band-structure effects in resonant scattering are governed by the terms $v_{\underline{G}}(z)$ with $\underline{G} \neq 0$.

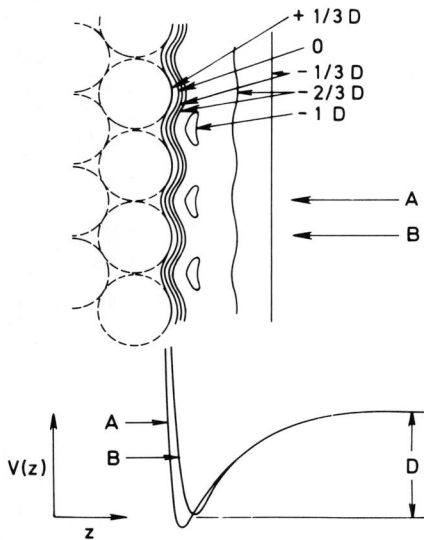

Fig.2.1. Schematic diagram illustrating the interaction of a neutral particle with a solid surface. The upper part shows equipotential lines and the negative potential energies are given in fractions of the depth of the potential. The lower part shows the potential as a function of z (normal to the surface) for two different lateral positions A and B /1.7/

2.2 Short Survey of Theoretical Efforts

As mentioned above, one method to obtain particle-surface potential functions $V(\underline{r})$ consists in the summation of individual particle-atom potentials $u(\underline{r})$,

$$V(\underline{r}) = \sum_{i} u(|\underline{r}-\underline{r}_{i}|) \quad , \tag{2.2}$$

with $\underline{r}_i$ denoting the location of the ith ion in the solid. This procedure is prob-
lematic for several reasons. Firstly, it requires knowledge of the location of all
ions in the solid. However, the atom positions near the surface might be shifted
from their corresponding bulk positions. Secondly, knowledge of the parameters describing the pair potential $u(\underline{r})$ is required. Except for rare-gas crystals, this
is a serious problem since the electronic state of the atoms in the solid is different from that in the gas phase, due to the chemical bond, and the particle-atom
interaction parameters are therefore different from those which can be obtained from
particle-particle scattering experiments /2.17/. Finally, many-body effects in the
solid, which certainly modify the pairwise interaction, are ignored. A discussion
of these problems can be found in the book by STEELE /2.18/. As an example, we cite
here results obtained by TSUCHIDA /2.19,20/ for He scattering from LiF(100). Assuming no relaxation of the surface ions, a Lennard-Jones type pair potential

$$u(r) = 4\varepsilon\left[\left(\frac{\sigma}{r}\right)^{12} - \left(\frac{\sigma}{r}\right)^{6}\right] \quad , \tag{2.3}$$

and replacing Li^{+} by He and F^{-} by Ne to estimate the interaction parameters ε and σ
(from scattering experiments in the gas phase), he found the surface potential depth
D and the bound-state energies ε_ν to be too large, whereas the equilibrium distance
was too small. Good agreement could be obtained only by modifying the parameters for
the $He-F^{-}$ interaction (the $He-Li^{+}$ interaction not being very important). Similar dif-
ficulties occurred in an investigation of the interaction of hydrogen atoms with
LiF(100) /2.4/. Further work of interest in this connection can be found in /2.21,
22/.

For large particle-surface separations (z larger than a typical lattice constant),
the asymptotic form of the attractive potential can be deduced from quite general
considerations /2.23,24/ and reads

$$V_{as}(z) = - C_3/z^3 \quad . \tag{2.4}$$

As the attractive potential is caused by the interaction of the particle with a large
number of crystal atoms, the asymptotic attraction part is practically independent of
$\underline{R}$, and the solid can be characterized by its complex dielectric response function
$\varepsilon(\omega)$. For an isotropic atom and an isotropic solid, C_3 can be written as /2.19/

$$C_3 = \frac{\hbar}{4\pi} \int_0^\infty \alpha(i\omega) \frac{\varepsilon(i\omega)-1}{\varepsilon(i\omega)+1} d\omega \quad , \tag{2.5}$$

where $\alpha(i\omega)$ is the electric polarizability of the particle and $\varepsilon(i\omega)$ is the dielectric function of the solid, both analytically continued to imaginary frequencies.

Table 2.1. Bound-state energies for different particle-surface systems as determined from resonant scattering data. Most of the data is taken from /1.10/

Particle	Surface	Bound-State Energies [meV]								Original Reference
		ε_0	ε_1	ε_2	ε_3	ε_4	ε_5	ε_6	ε_7	
H_1	LiF(100)	-12.3±0.3	-3.9±0.4	-0.5±0.5						/2.2/
	NaF(100)	-11.8±0.2	-3.0±0.2	-0.4±0.4						/2.2/
	KCl(100)-H_2O	-29.8±0.7	-22.5±0.6	-15.9±0.5	-10.3±0.3	-6.0±0.3				/2.3,4/
	Graphite (0001)	-31.6±0.2	-15.3±0.3							/2.5/
D_1	LiF(100)	-14.0±0.2	-6.7±0.2	-2.3±0.2	-0.5±0.3					/2.2/
	NaF(100)	-13.3±0.2	-5.8±0.2	-1.6±0.3	-0.3±0.3					/2.2/
	KCl(100)-H_2O			-19.3±0.5	-15.6±0.6	-11.0±0.4	-4.6±0.6			/2.4/
	Graphite (0001)	-35.4±0.2	-21.35±0.1	-12.0±0.1	-5.9±0.3					/2.5/
^3He	LiF(100)	-5.59±0.1	-2.00±0.1							/2.6/
	NaF(100)	-4.50±0.1	-1.38±0.1							/2.6,7/
	Graphite (0001)	-11.62±0.12	-5.38±0.12	-1.78±0.12						/2.8/

Table 2.1. (continued)

Particle	Surface	Bound-State Energies [meV]								Original Reference
		ε_0	ε_1	ε_2	ε_3	ε_4	ε_5	ε_6	ε_7	
^{4}He	LiF(100)	−5.9±0.1	−2.46±0.1	−0.78±0.1	−0.21±0.1					/2.6/
	NaF(100)	−4.92±0.1	−1.87±0.1	−0.54±0.1						/2.6,7/
	Graphite (0001)	−12.06±0.12	−6.36±0.12	−2.85±0.12	−1.01±0.12	−0.17±0.12				/2.8,9/
	NiO(100)	−7.8±0.4	−4.0±0.4	−1.6±0.4						/2.10,11/
	Cu(117)	−5.9±0.2	−4.0±0.3	−2.1±0.2						/2.12/
	Au(110) (1×2)	−6.0±0.3	−4.1±0.3	−2.2±0.3	−0.5±0.3					/2.13/ [a]
	Ni(110) + H(1×2)	−5.5	−2.6	−0.5						/2.14/ [a]
H$_2$	LiF(100)	−17.3±1.0	−10.0±1.0	−4.3±1.0						/2.15/
	Graphite (0001)	−41.61±0.25	−26.43±0.25	−15.33±0.25	−7.96±0.25	−3.61±0.25	−1.96±0.15			/2.16/
	NiO(100)	−48.0±2.0	−24.5±1.5	−19.5±1.5	−2.5±0.5	−3.7±0.5	−3.7±0.7			/2.10/
D$_2$	Graphite (0001)			−23.11±0.25	−15.4±0.25	−10.0±0.25	−6.37±0.25	−3.78±0.25	−1.93±0.25	/2.16/

[a] Preliminary analysis

Table 2.2. Values of potential-well depths D for different particle-surface combinations as calculated with Morse and 9-3 potential forms using the energy values of the two deepest levels ε_0 and ε_1 cited in Table 2.1 and averaged over isotopes. The next column gives results for D obtained by using (2.13) with $K_D = 12$ meV/10^{-25} cm³. Values of C_3, which rules the asymptotic behaviour of the attractive potential, as obtained from analysis of bound-state energy values /2.15/ as well as from calculations based on (2.11) are given in the last two columns. All data are from /1.10/ except the ones marked with an asterisk, which are from /2.26/

Particle	Surface	D[meV] Morse Potential	D[meV] 9-3 Potential	D[meV] Eq. (2.13)	C_3^{exp} [meV Å³]	C_3^{calc} [meV Å³]
³He, ⁴He	LiF(100)	- 8.10	- 8.70	- 7.93	120 ± 30	93
	NaF(100)	- 6.86	- 7.55	- 6.85	120 ± 40	73
	Graphite (0001)	- 15.55	- 16.3	- 14.9	110 ± 80	184
	NiO(100)	- 10.34	- 10.81	- 16.3	220 ± 110	-
H₁, D₁	LiF(100)	- 18.3	- 20.2	- 25.3	250 ± 90	194
	NaF(100)	- 18.6	- 20.9	- 21.8	180 ± 120	154
	Graphite (0001)	- 42.9	- 44.5	- 47.1	520*	397*
H₂	Graphite (0001)	- 50.5	- 51.6	- 57.5	544	573*
	NiO(100)	- 62.7	- 65.5	- 63.1	800 ± 190	-

Equation (2.5) has been used to calculate C_3 for H_1, D_1, and He interacting with LiF, NaF /2.25/, and graphite /2.26/. The results compare satisfactorily with results obtained from an analysis of experimental bound-state energies (Sect.2.3) as can be seen from Table 2.2.

Only a small number of attempts to establish from first principles the interaction of a noble gas atom with a metal surface can be found in the literature /2.27-30/. For such calculations, both the electronic structure of the atom as well as the electronic properties of the surface have to be taken into account. The electronic properties of surfaces are still a matter of intensive research /2.31/, so that necessary ingredients, like the charge density $n(\underline{r})$, needed to describe the short-range repulsive potential are known for only a few simple systems. ZAREMBA and KOHN /2.29/ have reported results for the interaction of He with surfaces of simple and noble metals using a jellium model (electron gas in a constant compensating background abruptly cutoff at the surface). The potentials derived are reproduced in Fig.2.2 for three noble metals in order to exhibit the potential depths and the equilibrium distance from the surface (jellium edge). It is clear that with a jellium model the lateral variation of the electron density is not taken into account, and therefore, only an approximation for the laterally averaged potential $v_{00}(z)$ is obtained. Unfortunately, no experimental results concerning bound-state energies of He on densely packed metal surfaces have been obtained up to now, so that a check of these theoretical results is not possible at present. It should be mentioned, however, that for the open metal surfaces Cu(117) /2.12/ and Au(110)(1×2) /2.13/, which have been investigated so far, the bound-state energies of the deepest levels are equally spaced in energy, so that the corresponding attractive potential is parabolic at the bottom and extends further in the z-direction than the potentials shown in Fig.2.2.

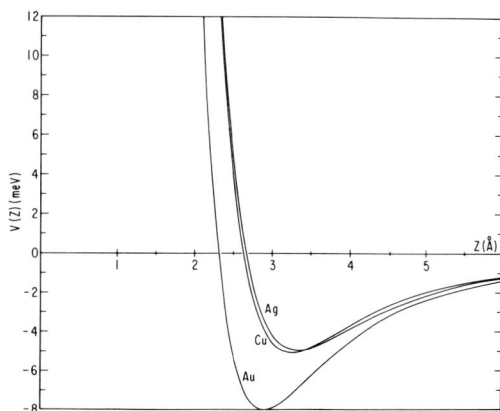

Fig.2.2. He-metal potentials for three noble metals as derived for jellium models by ZAREMBA and KOHN /2.24/. The origin corresponds to the edge of the jellium background

In a very recent study, ESBJERG and NORSKOV /2.32/ neglected the attractive part of the potential entirely and concentrated on the lateral variation of the repulsive part. They approximated the interaction energy $V(\underline{r})$ of a He atom with the electron-density profile $n(\underline{r})$ of a host system by replacing the inhomogeneous electron distribution of the host by an effective homogeneous medium with a density equal to the average of $n(\underline{r})$ over the electrostatic potential induced by a He atom immersed in that effective medium. They also showed that the calculated energy change in embedding a free He atom into a homogeneous electron gas is, to a good approximation, simply proportional to its density in the energy range of interest (20-100 meV). Using this approach, Esbjerg and Norskov showed that the He-He scattering potential, which is known from experiments, can be reproduced quite well and concluded that this local approximation should also be sufficient for a quantitative determination of the repulsive scattering potential of a surface since the density variations at a surface are of the same order of magnitude as in a He atom. A quantitative calculation was performed for the (110) surface of Al by adding a correction based on pseudo-potential theory to account for electron-density variations parallel to the surface. The resulting He equipotential curves along the [100] direction, which corresponds to the direction perpendicular to the close-packed Al rows, is exhibited in Fig.2.3.

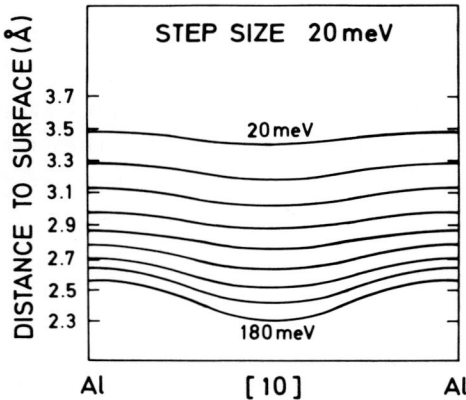

Fig.2.3. Equipotential lines for the repulsive part of the interaction of He with Al(110) in the direction perpendicular to the close-packed rows as calculated by ESBJERG and NORSKOV /2.32/. The distance perpendicular to the surface is measured from the first layer of Al atoms. In the direction of the close-packed rows the corrugation amplitude is not visible in the scale given

The equipotential profile $V(\underline{r}) = E$, which we shall call *corrugation function*, has a sinusoidal form and its amplitude increases slightly with the energy E of the He atom. Between 40 and 100 meV, the maximum amplitude ζ_m is $\cong 0.12$ Å. In the [$1\bar{1}0$] direction, the corrugation is much smaller (< 0.01 Å). Both results are in good agreement with experimental observations on clean metal surfaces (compare Sects.9.2,3). The important conclusion, which is worth emphasizing again, is that the repulsive He-scatter-

ing potential determined from scattering experiments is a direct probe of the surface electron density and therefore not only contains information on the geometrical
arrangement of the surface atoms, but also on the electronic structure of the surface.

2.3 Determination of the Surface Potential from Bound-State Energy Data

As mentioned in Sect.2.1 and discussed in detail in Sect.5.2, the discrete bound-
state energies can be determined in scattering experiments via the angular location
of resonant scattering features (Fig.5.2). The laterally averaged potential $v_{00}(z)$
is usually deduced by fitting the parameters of potentials described by simple ana-
lytic forms so as to reproduce the measured energy levels correctly. As an example,
in Fig.2.4 we show the interaction potential of deuterium with NaF(100) as determined by FINZEL et al. /2.2/. For $v_{00}(z)$, a Morse potential,

$$U(z) = D(e^{-2\sigma z} - 2e^{-\sigma z}) \quad , \tag{2.6}$$

with D denoting the depth and σ governing the range was used. For a particle of
mass m, the bound-state energies are given by

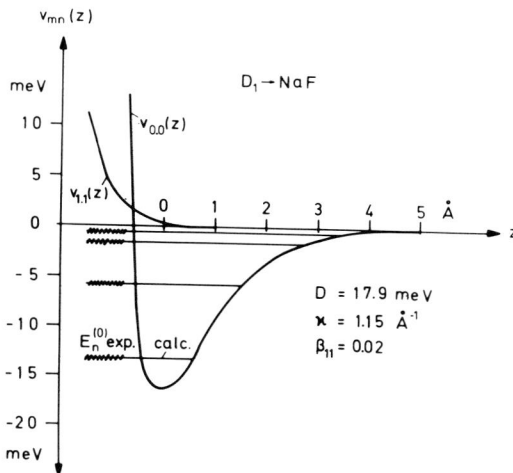

Fig.2.4. A Morse form for the laterally averaged surface potential $v_{00}(z)$ fits the
bound-state energies for D_1/NaF(100). The higher-order term $v_{11}(z)$ is also shown
/1.9/

$$\varepsilon_\nu = - \left(\frac{\sqrt{2mD}}{\sigma\hbar} - \nu - \frac{1}{2} \right) \frac{\sigma^2 \hbar^2}{2m} \qquad (2.7)$$

(h = 2πℏ denotes Planck's constant); ν can assume the values 0, 1, 2 ... and its maximum value is determined by the condition that the expression in the brackets must be positive. The four bound-state energies (Table 2.1) could be fitted within experimental uncertainty using (2.7) with D = 17.8 meV and σ = 1.15 Å^{-1}. It is of importance to note that with scattering of atomic hydrogen from NaF(100), exactly the same Morse-potential parameters D and σ were obtained /2.2/ although, due to the lighter mass of the H-atoms, only three different energy levels are allowed in this case [(2.4) and Table 2.1]. This proves that only the electronic properties of the particle and the solid surface determine the attractive part of the potential and also ensures that all bound states have been found.

Many other analytical forms for the laterally averaged potential $v_{00}(z)$ have been used in fitting the bound-state energy levels of the various systems investigated up to now (for a survey see /1.10/). We cite here another two, the first one being the 9-3 potential,

$$U(z) = \frac{D}{2} \left[\left(\frac{\sigma}{z} \right)^9 - \left(\frac{\sigma}{z} \right)^3 \right] \quad , \qquad (2.8)$$

which results from a pairwise summation of Lennard-Jones 12-6 potentials (2.3) in the continuum limit for the solid /2.19/. The second one is a very versatile three-parameter potential recently proposed by MATTERA et al. /2.33/, which can conveniently be used in analysing experimental data,

$$U(z) = D \left[\left(1 + \frac{\sigma z}{p} \right)^{-2p} - 2 \left(1 + \frac{\sigma z}{p} \right)^{-p} \right] \quad , \qquad (2.9)$$

with $- 1 \le 1/p \le 1$. D is again the depth of the potential; its width is determined by σ and its left-right symmetry depends on p. For a particle of mass m, the energy levels allowed are well approximated by

$$\varepsilon_\nu = - D \left[1 - \frac{\nu + 1/2 + [\delta(p)/A]}{AS(p)} \right]^{S(p)} \quad , \qquad (2.10)$$

where $A = \sqrt{2mD}/2\hbar\sigma$, $\delta(p) = (1 + 1/p)/32p$, and $S(p) = [1/2 - (1/4p)(3 + 1/p)]^{-1}$. The potential (2.5) encompasses many popular forms used for the description of $v_{00}(z)$, for example, the Morse potential (p → ∞), the harmonic potential (p = - 1), the Kratzer-Fues potential (p = 1), and the Lennard-Jones potential (2.3). It should be noted that not all of these potentials yield the correct asymptotic behaviour, (2.4).

As emphasized by LE ROY /2.15/, the energy-level spectrum $\{\varepsilon_\nu\}$ gives only a measure of the width of the potential well at the energy levels allowed, or in other words the distances between the corresponding classical turning points. Therefore, a given set of bound-state energies can be reproduced by a variety of model potentials with different minimum positions, depths, and shapes. For shallow energy levels, the potential should exhibit the correct asymptotic behaviour. This fixes the outermost turning point, so that the potential depth and asymptotic parameter C_3 can be determined by extrapolation /1.8,2.15/. Values for C_3 obtained in this way are listed in Table 2.2.

Values of D obtained by fitting the two deepest energy levels with both a Morse and a 9-3 potential are given in Table 2.2 for several particle-surface combinations. A comparison of both D values for each system illustrates the typical uncertainty to be expected in the determination of potential depths.

Recently, HOINKES and WILSCH /2.34/ have used the existing data to show that both the coefficients C_3 and the potential-well depth D can be estimated with relatively good accuracy ($\pm$ 15%) by using the relations

$$C_3 = K_C \alpha \frac{\varepsilon-1}{\varepsilon+1} \tag{2.11}$$

and

$$D = K_D \alpha \frac{\varepsilon-1}{\varepsilon+1} \quad , \tag{2.12}$$

where α denotes the static electric polarisability of the particle, ε the optical dielectric constant of the solid, and the coefficient K_C and K_D are

$$K_C = 140 \text{ meV } \text{Å}/10^{-25} \text{ cm}^3$$

and

$$K_D = 12 \text{ meV}/10^{-25} \text{ cm}^3 \quad .$$

For comparison with D values obtained from fitting bound-state energies, Table 2.2 also gives values obtained from (2.12).

The dominant $\underline{G} \neq 0$ potential terms $v_G(z)$ can be determined by investigating characteristic band structure splittings observed in resonant scattering (Sect.5.2). This has been done by HOINKES et al. /2.3/ for the system $D_1/NaF(100)$, for which the laterally averaged potential $v_{00}(z)$ can be described by a Morse potential, (2.6). As the periodic surface structure effects mainly the short-range repulsion, the higher potential terms were assumed to be only repulsive,

$$v_{\underline{G}}(z) = \beta_{\underline{G}} \, D \, e^{-2\kappa z} \quad . \tag{2.13}$$

Both the v_{10} and v_{11} terms are of importance for D_1/NaF, and $\beta_{11} \sim \beta_{10}/2$; the contribution due to $v_{11}(z)$ is shown in Fig.2.4.

A very interesting recent contribution which should be mentioned in the context of determinations of potentials from experimental data is that of CARLOS and COLE /2.35/ for He on the basal plane of graphite. This system has been investigated in great detail experimentally /2.8,9,36,37/, and both the bound-state energies (see Table 2.1) as well as the band-splitting parameters (only v_{10} being significant /2.9/) have been determined with great accuracy. Carlos and Cole started from the assumption that the gas-surface interaction $V(\underline{r})$ can be represented by the sum of pairwise potentials $u(\underline{r})$ [compare (2.2)]. In order to obtain optimum agreement between the measured set of bound-state energies (for both ^{4}He and ^{3}He) and the band-splitting parameters, they evaluated a variety of model pair potentials and arrived at the conclusion that all isotropic pair potentials underestimate the value of v_{10}. The introduction of anisotropic pair potentials, which can be justified on the basis of the anisotropic electric properties of graphite, leads to a much better agreement. Two anisotropic pair potentials (12-6 and Yukawa-6 forms) give equivalent results for the net interaction $V(\underline{r})$; they are shown in Fig.2.5. As one can observe, the anisotropy is such that the attractive Van der Waals' force is weakest when the He is above a C atom. The functions $V(\underline{r})$ resulting from the two different pair potentials are shown in Fig.2.6. In Fig.2.7, the corresponding equipotential curves are reproduced. The He/graphite potential determined in this way allowed calculation of thermodynamic data which are in impressive agreement with experimental results /2.38/.

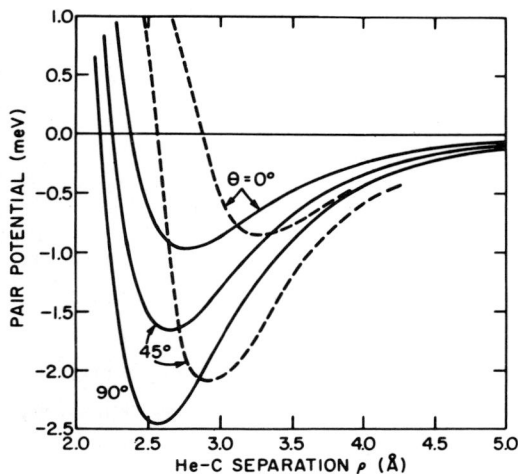

Fig.2.5. Anisotropic pair potential $u(r,\theta)$ as derived from CARLOS and COLE /2.35/ for He/graphite. r denotes the He-carbon separation and θ is the angle between the surface normal and the He-carbon connecting line. (---): 12-6 interaction; (——): Yukawa-6 interaction

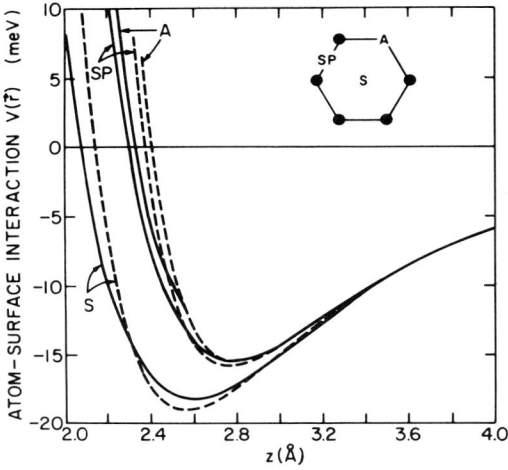

Fig.2.6. He/graphite surface interaction $V(\underline{r})$ as a function of z above symmetry points in the basal plane indicated in the upper right. (---): anisotropic 12-6 pair interaction; (——): anisotropic Yukawa-6 pair interaction (for these curves the z values have been increased by 0.5 Å) /2.35/

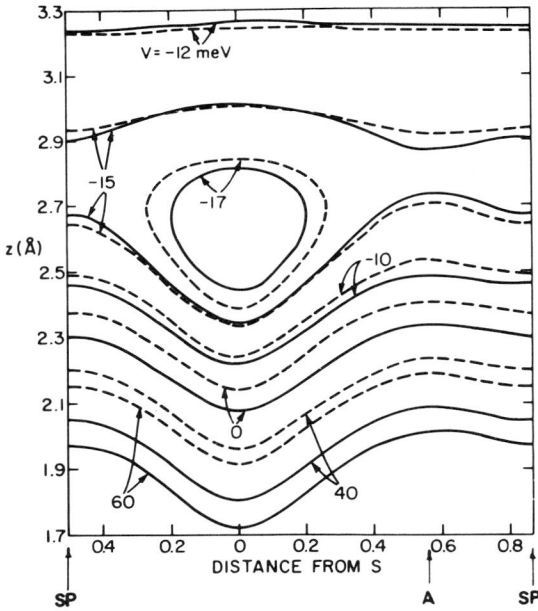

Fig.2.7. Equipotential lines in meV for the He-graphite potential. The abscissa denotes the distance in the basal plane from point S (Fig.2.6) in units of the separation between centres of neighbouring hexagons on graphite. (---): anisotropic 12-6 pair potentials; (——): anisotropic Yukawa-6 potential (z is increased for the full curves by 0.5 Å) /2.35/

In closing this chapter, we want to emphasize that from scattering experiments, information on different parts of the particle-surface potential $V(\underline{r})$ is obtained (see Fig.2.6):

a) The intensities of the diffraction peaks are primarily determined by the periodic modulation of the repulsive part of the potential described by the corrugation function $\zeta(R)$. The shape and maximum amplitude of $\zeta(R)$ depends slightly on the energy of the incoming particles as their classical turning points are closer to the cores of the surface ions or atoms when the particles are faster (compare Fig.2.3). For typical energies of the incoming particles, this region corresponds to potential energies $V \gg D$ and $z \ll \sigma$. The corrugation function is essentially a replica of a contour of constant valence-electron density above the top surface layer and contains the essential structural information. Its derivation from experimental data is the subject of Chap.3, and the currently available structural information based on corrugation functions is surveyed in Chaps.7-10.

b) As discussed in detail in Sect.5.2, the analysis of resonant scattering structures yields the bound-state energies, and as outlined above, knowledge of the bound-state energies allows reconstruction of the laterally averaged low-energy part of the potential $v_{00}(z)$, whereby the correct asymptotic behaviour (2.4) should be taken into account. Thus, information for the region $V < 0$ is obtained.

c) Analysis of band-splitting effects in resonant features (Sect.5.2) allows determination of the $G \neq 0$ Fourier coefficients $v_G(z)$, which govern the periodic modulation parallel to the surface of the attractive-well minimum as well as the modulation of the repulsive part in this low-energy range (see Fig.2.4).

It is clear that a complete description of the particle-surface interaction potential has to match these different parts. In the above-mentioned determination of the He/graphite potential from bound-state resonances and band splittings, extrapolation to the energy range $E \geq 0$ yields for both anisotropic pair potentials a maximum corrugation between the points A and S in the unit cell (Fig.2.6) of 0.26 Å at $E = 0$ meV and of 0.29 Å at $E = 60$ meV. An intensity analysis of Bragg peaks for He energies of 63 meV has given a maximum corrugation of 0.21 Å. (Note that a different distance of the corrugation from the ion cores does not influence the diffraction intensities.) A similar discrepancy is found for He/LiF, and is discussed in Chap.7. This discrepancy has been attributed /2.35/ to the assumption of a corrugated hard wall in analyzing the diffraction intensities (Chap.3). However, as pointed out in Sect.3.8, the softness of a repulsive potential will not appreciably influence corrugations with rather large amplitudes. It appears more probable that the discrepancies in question arise because the assumed simple analytical forms of the repulsive part of the pair potentials $u(\underline{r})$ are unable to properly describe the electron-density variation at a surface. Unfortunately, to date no experimental results are available on the variation of $\zeta(\underline{R})$ as a function of energy which might shed further light onto this problem.

3. Quantum Theory of Particle Diffraction

3.1 The Corrugated Hard-Wall Model

Diffraction patterns, in general, contain two kinds of information: the dimensions of the unit cell determine the angular location of the diffraction peaks, and the spatial distribution of the scattering centres determines their intensities. In this chapter, we outline the mathematical formalism describing molecular-beam diffraction, give formulae to calculate the angular location of diffraction peaks, and discuss procedures to calculate their intensities. For the latter purpose, we make three simplifying assumptions to obtain a mathematically convenient and physically reasonable description of the particle-surface potential. Firstly, we neglect the attractive part of the potential. Using incoming particle energies of $E_i \sim 60$ meV, this is a reasonable assumption, as with potential depths of the order of $D \sim 10$ meV (compare Table 2.2) the effective energy $E_i + D$ with which the particles "hit" the repulsive wall is not very different from E_i. For E_i comparable to D, simple corrections can be applied (Sect.5.1). Secondly, the steeply rising repulsive part of the potential is regarded as having infinite slope. Historically, this assumption was put forward by Lord RAYLEIGH /3.1/ in his investigations concerning the scattering of sound waves from walls. For the quantum theory of atomic-beam scattering, it was introduced by GARIBALDI et al. /3.2/. Thirdly, the influence of lattice vibrational modes which give rise to a periodic modulation in time of the potential is neglected and the repulsive part of the potential is therefore equivalent to that of a rigid hard wall.

   With these assumptions, the particle-surface interaction potential can be represented as

$$V(z) = 0 \text{ for } z > \zeta(\underline{R})$$
$$V(z) = \infty \text{ for } z \leq \zeta(\underline{R})$$
(3.1)

with z denoting the direction of the surface normal. The full crystallographic information obtainable from a diffraction experiment with atomic beams is contained in the two-dimensional corrugation function $\zeta(\underline{R}) = \zeta(x,y)$ with periods $\underline{a}_1$ and $\underline{a}_2$. As outlined in Chap.2, $\zeta(\underline{R})$ is a replica of the spatial modulation of the valence electron density of the scattering surface. Due to the fact that in reality the repulsive part of the potential does not have infinite slope, $\zeta(\underline{R})$ is in principle a function of the energy E_i of the incoming particles. However, in the range of energies applicable in actual diffraction experiments ($E_i \sim 20$-100 meV), both the maximum amplitude ζ_m of $\zeta(\underline{R})$ as well as its shape will be affected only slightly (Sects.2.2,3.9).

Due to its two-dimensional periodicity, the "hard-wall corrugation function" $\zeta(\underline{R})$ can be conveniently written as a Fourier series

$$\zeta(\underline{R}) = \sum_{\underline{G}} \zeta_{\underline{G}} \, e^{i\underline{G}\underline{R}} \quad , \tag{3.2}$$

where the reciprocal lattice vectors $\underline{G} = j\underline{b}_1 + \ell\underline{b}_2$ $(j,\ell = 0, \pm 1, \pm 2 \dots)$ are related to the unit-cell vectors $\underline{a}_1$ and $\underline{a}_2$ of the direct lattice through

$$\underline{a}_p\underline{b}_q = 2\pi\delta_{pq} \tag{3.3}$$

with δ_{pq} denoting the Kronecker symbol. Equation (3.3) implies that $\underline{b}_1$ and $\underline{b}_2$ are normal to $\underline{a}_2$ and $\underline{a}_1$, respectively. With β being the angle between $\underline{a}_1$ and $\underline{a}_2$, the lengths of $\underline{b}_1$ and $\underline{b}_2$ are determined by

$$|\underline{b}_1| = \frac{2\pi}{a_1\sin\beta} \qquad |\underline{b}_2| = \frac{2\pi}{a_2\sin\beta} \quad . \tag{3.4}$$

The relation between direct and reciprocal lattices is illustrated in Fig.3.1.

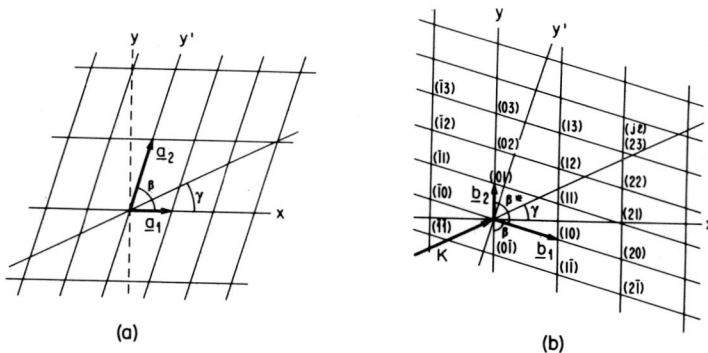

(a)

(b)

<u>Fig.3.1a,b.</u> Graphical representation of the relation between a surface lattice (a) and the corresponding reciprocal lattice (b) according to (3.3,4)

3.2 Diffraction Condition and Ewald Construction

Consider a beam of particles with energy E_i impinging on a surface at an angle θ_i as measured from the surface normal. The particle energy E_i is related to the beam wavelength λ_i according to the de Broglie relation

$$\lambda_i = \frac{h}{\sqrt{2mE_i}} \tag{3.5}$$

with m denoting the mass of the particles, and h being Planck's constant. The wave-vector $\underline{k}_i$ is determined by the wavelength and the direction of the beam

$$k_i \equiv |\underline{k}_i| = \frac{2\pi}{\lambda_i} \quad . \tag{3.6}$$

We separate $\underline{k}_i$ into components parallel and perpendicular to the surface:

$$\underline{k}_i \equiv (\underline{K}, k_{iz}) = (k_i \sin\theta_i, - k_i \cos\theta_i) \quad . \tag{3.7}$$

The well-known Bragg condition for diffraction to occur is

$$\underline{K} + \underline{G} = \underline{K}_G \quad . \tag{3.8}$$

In the diffraction process, the total energy of the particles remains unchanged,

$$\underline{k}_i^2 = \underline{k}_G^2 \quad . \tag{3.9}$$

In writing (3.7), we have again separated the wavevectors for the diffracted beams into parallel and perpendicular components:

$$\underline{k}_G = (\underline{K}_G, k_{Gz}) = (k_i \sin\theta_G, k_i \cos\theta_G) \quad . \tag{3.10}$$

Equation (3.9) restricts the number of reciprocal lattice vectors for which diffraction can occur to the finite set $\{\underline{F}\}$ for which

$$k_{Fz}^2 = k_i^2 - (\underline{K} + \underline{F})^2 > 0 \quad . \tag{3.11}$$

In the case of a one-dimensional corrugation function $\zeta(x)$ with period a for which the $\underline{G}$ vectors are given by $G = j(2\pi/a)$ $(j = 0, \pm 1, \pm 2)$, (3.8) yields

$$\sin\theta_j = \sin\theta_i + j\frac{\lambda_i}{a} \quad . \tag{3.12}$$

The conditions (3.7,8) can be represented graphically with the Ewald construction shown for the example of a one-dimensional surface in Fig.3.2.

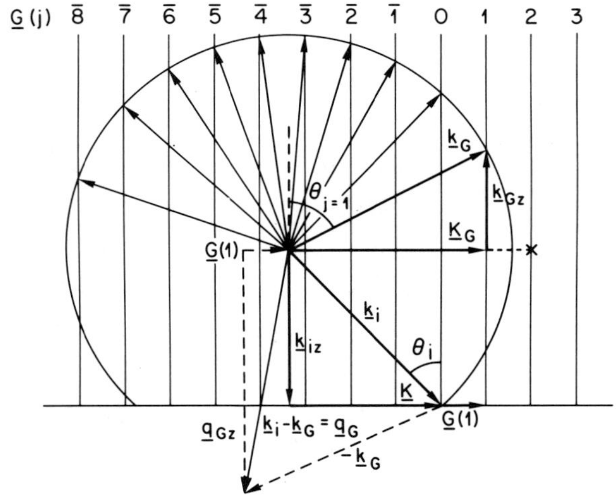

Fig.3.2. The Ewald construction for diffraction from surfaces

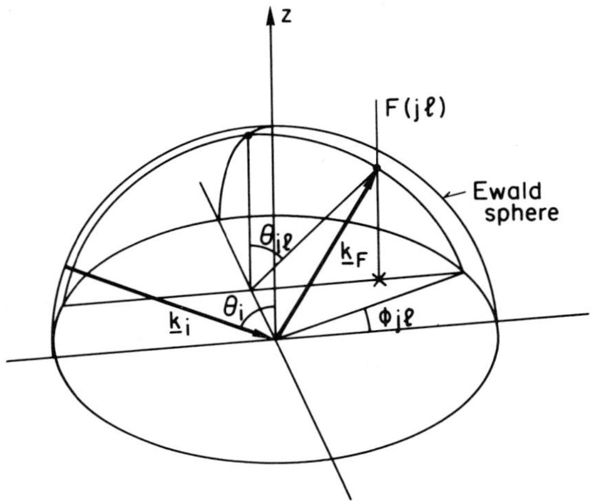

Fig.3.3. Specification of the diffraction angles θ_G and ϕ_G in accordance with possible motions of a goniometer (see Fig.6.11)

For two-dimensional corrugation functions $\zeta(x,y)$, two angles specifying the scattering direction have to be determined. According to the motions usually possible with a goniometer (Fig.6.11), we choose these angles in the following manner (Fig. 3.3). θ_G describes the angle by which the detector has to be rotated in the scattering plane, which is spanned by the wave vector $\underline{k}_i$ and the surface normal; θ_G is measured from the surface normal. ϕ_G denotes the angle through which the detector has to be moved out of plane. For an arbitrary two-dimensional lattice with unit-cell vectors $\underline{a}_1$ and $\underline{a}_2$ comprising the angle β, θ_G and ϕ_G can be calculated from the following formulae ($\underline{G} = j\underline{b}_1 + \ell\underline{b}_2$):

$$\sin\phi_{j\ell} = \left[-\frac{j}{a_1}(\sin\gamma + \cos\gamma \cot\beta) + \frac{\ell}{a_2}\frac{\cos\gamma}{\sin\beta} \right]\lambda_i \qquad (3.13a)$$

$$\sin\theta_{j\ell} = \frac{1}{\cos\phi_{j\ell}} \left[\sin\theta_i + \lambda_i \frac{j}{a_1} (\cos\gamma - \sin\gamma \cot\beta) + \lambda_i \frac{\ell}{a_2} \frac{\sin\gamma}{\sin\beta} \right] \quad . \qquad (3.13b)$$

In these equations, γ denotes the angle between $\underline{K}$ and $\underline{a}_1$ (Fig.3.1a).

3.3 Calculation of Diffraction Intensities — General Method

A method to calculate diffraction intensities on a quantum-mechanical basis for general hard corrugated surfaces $\zeta(\underline{R})$ and general scattering geometries has been developed in the last few years as a result of the effort of several groups /3.3-9/.
The starting point is the Lippman-Schwinger equation, which can be solved for the
hard corrugated potential (3.1), yielding for the wave function of the incoming and
scattered particles

$$\psi(\underline{r}) = \exp[i(\underline{KR}+k_{iz}z)] + \int_{\substack{unit \\ cell}} d\underline{R}'f(\underline{R}') \sum_{\underline{G}} \frac{1}{k_{\underline{G}z}} \exp[i(\underline{K}+\underline{G})(\underline{R}-\underline{R}')] \exp[ik_{\underline{G}z}|z-\zeta(\underline{R})|]$$

$$(3.14)$$

with $f(\underline{R})$ denoting the density of sources.

It should be mentioned that the numerical value of k_{iz} is negative (3.7), and
$k_{0z} = - k_{iz}$, with k_{0z} denoting the z-component of the wave vector of the specular
beam $\underline{G} = 0$. The far-field solution, obtained with $z \to \infty$,

$$\psi(\underline{r}) = \exp[i(\underline{KR}+k_{iz}z)] + \sum_{\underline{G}} A_{\underline{G}} \exp[i(\underline{K}+\underline{G})\underline{R}] \exp[ik_{\underline{G}z}z] \qquad (3.15)$$

contains the (complex) scattering amplitudes $A_{\underline{G}}$ which determine the scattered intensities $P_{\underline{G}}$ through

$$P_{\underline{G}} = \frac{|k_{\underline{G}z}|}{|k_{iz}|} |A_{\underline{G}}|^2 \quad . \qquad (3.16)$$

As all the scattering from the rigid hard wall is elastic, the diffracted intensities have to satisfy the unitarity condition

$$\sum_{\underline{F}} P_{\underline{F}} = 1 \quad . \qquad (3.17)$$

Comparison of (3.14) and (3.15) yields for the scattering amplitudes

$$A_{\underline{G}} = \frac{1}{k_{\underline{G}z}} \int d\underline{R}'f(\underline{R}') \exp[-i(\underline{K}+\underline{G})\underline{R}'] \exp[-ik_{\underline{G}z}\zeta(\underline{R}')] \quad . \qquad (3.18)$$

To be able to calculate A_G from (3.18), knowledge of the source function $f(\underline{R})$ is required. With the boundary condition

$$\psi[\underline{R}, z = \zeta(\underline{R})] \equiv 0 \quad , \tag{3.19}$$

which simply expresses the fact that the particles cannot penetrate the surface, we obtain from (3.14),

$$- \exp[ik_{iz}\zeta(\underline{R})] = \sum_{\underline{G}} \frac{1}{k_{Gz}} \exp[i\underline{GR}] \int_{\substack{\text{unit} \\ \text{cell}}} d\underline{R}'f(\underline{R}') \exp[-i(\underline{K}+\underline{G})\underline{R}'] \exp[ik_{Gz}|\zeta(\underline{R})-\zeta(\underline{R}')|] \tag{3.20}$$

This equation has to be solved to obtain the source function $f(\underline{R})$. GARCIA and CABRERA /3.7,8/ have proposed a numerical procedure to deduce $f(\underline{R})$ from (3.20) called the RR' method, which we shall sketch in the following. We first multiply both sides of (3.20) by $\exp(i\underline{KR})$ and exchange summation and integration on the right-hand side:

$$- \exp\{i[\underline{KR}+k_{iz}\zeta(\underline{R})]\} = \int d\underline{R}'f(\underline{R}') \sum_{\underline{G}} \frac{\exp[i(\underline{K}+\underline{G})(\underline{R}-\underline{R}')]}{k_{Gz}} \exp[ik_{Gz}|\zeta(\underline{R})-\zeta(\underline{R}')|] \tag{3.21}$$

Now, we substitute the integration by a summation and for simplicity restrict ourselves to a one-dimensional corrugation $\zeta(x)$. Taking 2N equidistant points $x_n = na/2N$ within the length a of the elementary cell, we obtain a set of 2N linear equations which has to be solved for the 2N unknowns $f_{n'} \equiv f(x_{n'})a/2N$,

$$B_n = \sum_{n'=1}^{2N} f_{n'}M_{nn'} \quad , \tag{3.22}$$

whereby the B_n and $M_{nn'}$ are given by

$$B_n = \exp\{i[Kna/2N+k_{iz}\zeta(x_n)]\} \tag{3.23}$$

and

$$M_{nn'} = \sum_{\underline{G}} \frac{1}{k_{Gz}} \exp[i(K+G)(n-n')a/2N] \exp[ik_{Gz}|\zeta(x_n)-\zeta(x_{n'})|] \quad . \tag{3.24}$$

For calculation of the $M_{nn'}$, the summation has to be performed over a sufficient set of $\underline{G}$ vectors so that $M_{nn'}$ is "close enough" to the value for $\underline{G} \to \infty$. This can be checked numerically /3.8/. For n = n', (3.22) cannot be used, as the summation over the $\underline{G}$'s diverges. GARCIA and CABRERA /3.7,8/ have derived an analytic expression using the linear approximation $|\zeta(x_n) - \zeta(x_{n'})| \cong |\zeta'(x_n)||x_n - x_{n'}|$, which should read

$$M_{nn'} = \sum_{\underline{G}} \frac{2N}{k_{\underline{G}z}a} \frac{\exp\{i\left[(\underline{K}+\underline{G})+|\zeta'(x_n)|k_{\underline{G}z}\right]a/4N\}-1}{i\left[(\underline{K}+\underline{G})+|\zeta'(x_n)|k_{\underline{G}z}\right]}$$

$$+ \frac{1-\exp\{-i\left[(\underline{K}+\underline{G})-|\zeta'(x_n)|k_{\underline{G}z}\right]a/4N\}}{i\left[(\underline{K}+\underline{G})-|\zeta'(x_n)|k_{\underline{G}z}\right]} \quad .$$

(3.25)

With (3.22) solved for the f_n, the scattering amplitudes $A_{\underline{G}}$ are obtained by using (3.16) in discretized form,

$$A_{\underline{G}} = \frac{1}{k_{\underline{G}z}} \sum_{n=1}^{2N} f_n \exp\left[-i(\underline{K}+\underline{G})an/2N\right] \exp\left[-ik_{\underline{G}z}\zeta(x_n)\right] \quad .$$

(3.26)

Model calculations based on this numerical procedure for one-dimensional corrugations of different shape and amplitude seem to confirm that the method is applicable to any kind of corrugation with no restriction concerning the maximum corrugation amplitude or scattering geometry $(\underline{k}_i,\theta_i,\gamma)$. Even corrugations with discontinuities in $\zeta(x)$ and/or $d\zeta(x)/dx$ can be treated /3.10/. Although such corrugations are highly un-likely to play a role in atomic-beam scattering, the same formalism also applies to other problems, for example, the scattering of acoustic waves from solid walls. Examples for the spatial variation of the real and imaginary parts of the source function $f(x)$ for differently shaped corrugations $\zeta(x)$ with increasing amplitude are given in /3.8/. It is worthwhile mentioning that for small corrugation ampli-tudes, the real part of $f(x)$ is nearly constant with a value of $\sim k_{iz}$, and the ima-ginary part is very small.

The solution of (3.22) requires handling of very large matrices, and even for the one-dimensional model corrugations considered in /3.7-10/ the number 2N had to be taken between 100 and 200. As a consequence, although the method is in principle applicable to two-dimensional corrugations $\zeta(x,y)$, up to now no such calculation has been reported. The reliability of the calculated intensities is usually judged using the following two criteria. a) The unitarity relation (3.17) has to be veri-fied within a fraction of a percent. b) The calculation must yield the correct threshold behaviour of the intensities: the intensity $P_{\underline{F}}$ of a beam $\underline{F}$ disappearing at the horizon $(\theta_F = 90°, k_{Fz} = 0)$ must decrease to zero with a vertical tangent as $\underline{k}_i$, θ_i, or γ vary so that θ_F approaches $90°$ /3.8,11/. (This behaviour can be observed in the model calculations of Figs.3.7,8.)

3.4 Calculation of Intensities — Rayleigh Hypothesis

An appreciable simplification of the calculational procedure to obtain the scattering amplitudes A_G is achieved by an assumption put forward almost a century ago in Lord RAYLEIGH's first theoretical investigations of the reflection of acoustic waves from hard walls /3.1/. Rayleigh assumed that the far-field solution (3.15) is strictly valid all the way to the surface. Imposing the boundary condition (3.19) on (3.15), we obtain

$$\exp\{i[\underline{KR}+k_{iz}\zeta(\underline{R})]\} + \sum_{\underline{G}} A_{\underline{G}} \exp\{i[\underline{KR}+\underline{GR}+k_{Gz}\zeta(\underline{R})]\} = 0 \tag{3.27}$$

or

$$\sum_{\underline{G}} A_{\underline{G}} \exp[ik_{Gz}\zeta(\underline{R})] \exp[i\underline{GR}] = -\exp[ik_{iz}\zeta(\underline{R})] \quad . \tag{3.28}$$

These equations must be fulfilled for every point $\underline{R}$ at the surface (the two-dimensional periodicity permits restriction to a single surface unit cell in actual calculations) and allows direct determination of the A_G's without the necessity of knowing the source function $f(\underline{R})$ explicitly. Of course, there is a price to pay for this simplifying Rayleigh assumption, which consists in the limited convergence range of (3.28). The limits of convergence were investigated analytically in /3.12-15/. For a one-dimensional sinusoidal corrugation with periodicity a,

$$\zeta(x) = \frac{1}{2} \zeta_m \cos \frac{2\pi}{a} x \quad , \tag{3.29a}$$

the Rayleigh method converges up to

$$\zeta_m = 0.143\ a \quad . \tag{3.29b}$$

For a two-dimensional quadratic corrugation with lattice constant a,

$$\zeta(x,y) = \frac{1}{4} \zeta_m \left(\cos \frac{2\pi}{a} x + \cos \frac{2\pi}{a} y \right) \quad , \tag{3.30a}$$

the limit of convergence is given by

$$\zeta_m = 0.188\ a \quad . \tag{3.30b}$$

Both (3.29) and (3.30) are written in such a way that ζ_m denotes the maximum corrugation amplitude. VAN DEN BERG and FOKKEMA /3.15/ have investigated finite Fourier series of triangular and rectangular one-dimensional corrugation profiles and have shown that the convergence limits become smaller as the number of terms increases, or in other words, as the smallest radius of curvature present at the surface approaches zero. Consequently, the Rayleigh approach breaks down for corrugations exhibiting discontinuities in $\zeta(\underline{R})$ and/or $d\zeta(\underline{R})/d\underline{R}$.

3.4.1 The GR Method

This very powerful procedure to compute the $A_{\underline{G}}$'s from (3.28) was developed by GARCIA et al. /3.16-19/. Rearranging (3.28) by multiplying each side by $\exp[-ik_{iz}\zeta(\underline{R})]$, we obtain

$$\sum_{\underline{G}} A_{\underline{G}} M_{\underline{G}\underline{R}} = -1 \tag{3.31}$$

with

$$M_{\underline{G}\underline{R}} = \exp\{i[(k_{\underline{G}z}-k_{iz})\zeta(\underline{R})+\underline{G}\underline{R}]\} \quad . \tag{3.32}$$

Equation (3.32) must be satisfied for every point $\underline{R}$ in the unit cell. If one now chooses a finite set of n vectors $\underline{R}_n$ uniformly distributed over the surface unit cell and relates them to the same number of uniformly distributed reciprocal lattice vectors $\underline{G}$, one can regard (3.31) as a set of n linear equations which can be solved for the $A_{\underline{G}}$'s. The computational applicability of this method was proven by GARCIA /3.19/ in many one- and two-dimensional model calculations, and especially in the first surface crystallographic investigation using He diffraction /3.18/, which was based on the experimental data of BOATO et al. /3.20/, yielding the corrugation function for the (100) surface of LiF (Chap.7). Garcia also showed that reliable numerical results could be obtained for maximum corrugation amplitudes slightly above the analytical convergence limits given above. Depending on the form of the corrugation, its maximum amplitude, and the scattering conditions $(\underline{k}_i,\theta_i)$, the dimension n of the matrix $M_{\underline{G}\underline{R}}$ may be very large (in his LiF study, Garcia used n between 100 and 200) and the calculations may therefore be rather time-consuming.

3.4.2 The Eikonal Approximation

Starting again from (3.28), we multiply both its sides by $\exp[-i(\underline{G}'\underline{R}+k_{\underline{G}'z}\zeta(\underline{R})]$, then integrate over the unit cell and obtain

$$\sum M_{\underline{GG}'} A_{\underline{G}} = A^0_{\underline{G}'}$$ (3.33)

with

$$M_{\underline{GG}'} = \frac{1}{\mathscr{F}} \int \exp\left[i(\underline{G}-\underline{G}')\underline{R}+i(k_{\underline{G}z}-k_{\underline{G}'z})\zeta(\underline{R})\right] d\underline{R}$$ (3.34)

and

$$A^0_{\underline{G}} = -\frac{1}{\mathscr{F}} \int \exp\{-i[\underline{G}\underline{R}+(k_{\underline{G}z}-k_{iz})\zeta(\underline{R})]\} d\underline{R} \quad ,$$ (3.35)

where $\mathscr{F}$ denotes the unit-cell area. Restricting the possible $\underline{G}$'s to a finite set of n vectors, (3.33) yields a set of n linear equations which can be solved for the $A_{\underline{G}}$'s. However, as values of n of the same magnitude have to be used in the case of the GR method /3.21/, actual calculations using this formulation are even more time-consuming, because in addition to handling the large matrix $M_{\underline{GG}'}$, the integrals (3.34,35) also have to be evaluated with good accuracy. Nevertheless, from (3.33) one can readily derive an important approximation /3.2/, which is very convenient to use for calculations of scattering amplitudes. As one can see from (3.34), the diagonal elements of the matrix $M_{\underline{GG}'}$ are all unity. Under certain conditions, the out-of-diagonal elements will be small and can be neglected so that (3.33) becomes

$$A_{\underline{G}} = A^0_{\underline{G}} \quad .$$ (3.36)

This is the so-called eikonal approximation, from which the diffraction probabilities can readily be calculated by simply evaluating the integral (3.35). The conditions under which (3.35) can be used with satisfactory accuracy are the following. The corrugation function $\zeta(\underline{R})$ has to be smooth, and its maximum amplitude should be small compared to the lattice constant ($\zeta_m \leq 0.1a$). Furthermore, the angle of incidence must be small, so that all intense diffraction beams appear far from grazing emergence. The latter condition is a consequence of the fact that neglect of the non-diagonal terms in (3.33) corresponds to a neglect of the contribution of evanescent waves ($k_{\underline{G}z}^2 < 0$), which belong to $\underline{G}$ vectors outside the Ewald sphere. As a consequence, the eikonal approximation fails to describe correctly the threshold behaviour of the intensities of the beams near horizon. The classical analogue ($\lambda_i \to 0$) for the condition that evanescent waves do not play a role is that multiple collisions of the particles with the hard wall do not occur.

The unimportance of evanescent waves has a remarkable nontrivial consequence for surface structural investigations using atomic-beam diffraction: the diffraction intensities of $\zeta(\underline{R})$ and $-\zeta(-\underline{R})$ are the same as long as coupling to evanescent waves is negligible /3.22/. For surfaces with two-dimensional inversion symmetry,

$\zeta(\underline{R}) = \zeta(-\underline{R})$, which form the overwhelming number of real surfaces, this means that from an analysis of diffraction intensities under conditions where the eikonal approximation works well, one cannot decide whether $+\zeta(\underline{R})$ or $-\zeta(R)$ describes the surface profile. To prove this statement, we consider (3.31), which for $+\zeta(\underline{R})$ reads in full form

$$\sum_{\underline{G}} A_{\underline{G}} \exp\{i[(k_{\underline{G}z}-k_{iz})\zeta(\underline{R})+\underline{GR}]\} = -1 \quad . \tag{3.37}$$

With no contribution of evanescent waves, all the $k_{\underline{G}z}$ are real. An analogous equation holds for $-\zeta(-\underline{R})$; for this case we denote the scattering amplitudes by $\overline{A}_{\underline{G}}$,

$$\sum_{\underline{G}} \overline{A}_{\underline{G}} \exp\{i[-(k_{\underline{G}z}-k_{iz})\zeta(-\underline{R})+\underline{GR}]\} = -1 \tag{3.38}$$

We can take the complex conjugate of (3.37) to obtain the equivalent equation for $A_{\underline{G}}^*$. Furthermore, as this equation has to be fulfilled at every point of the surface, we can perform the transformation $\underline{R} \rightarrow -\underline{R}$ without changing its validity. Combining both steps, we obtain

$$\sum_{\underline{G}} A_{\underline{G}}^* \exp\{i[-(k_{\underline{G}z}-k_{iz})\zeta(-\underline{R})+\underline{GR}]\} = -1 \quad . \tag{3.39}$$

Comparison of (3.38,39) shows that $\overline{A}_{\underline{G}} = A_{\underline{G}}^*$. Therefore, for both $+\zeta(\underline{R})$ and $-\zeta(-\underline{R})$, the diffraction intensities $P_{\underline{G}} \sim |A_{\underline{G}}|^2 = |\overline{A}_{\underline{G}}|^2 = A_{\underline{G}}A_{\underline{G}}^*$ are the same.

This proof does not hold for evanescent waves. Indeed, writing $k_{\underline{G}z} = i\chi_{\underline{G}z}$ in any of the terms of (3.37,38), which correspond to evanescent waves, one easily verifies that complex conjugation plus inversion yields for (3.37)

$$A_{\underline{G}}^* \exp[-(ik_{iz}+\chi_{\underline{G}z})\zeta(\underline{R})+i\underline{GR}] \quad ,$$

whereas the analogous term in (3.38) yields

$$\overline{A}_{\underline{G}} \exp[-(ik_{iz}-\chi_{\underline{G}z})\zeta(\underline{R})+i\underline{GR}] \quad .$$

Therefore, opposite corrugations will be impossible to distinguish in practice if the contribution of evanescent waves is small or, in other words, if multiple scattering is negligible. We illustrate this in Fig.3.4, using the example of classical elastic scattering of particles from a simple one-dimensional hard-wall model corrugation with inversion symmetry $\zeta(\underline{R}) = \zeta(-\underline{R})$. For the case of classical scattering, the trajectories of the particles can be traced. In Fig.3.4a, the maximum corrugation is small, and with the particle beam impinging at normal incidence, three dis-

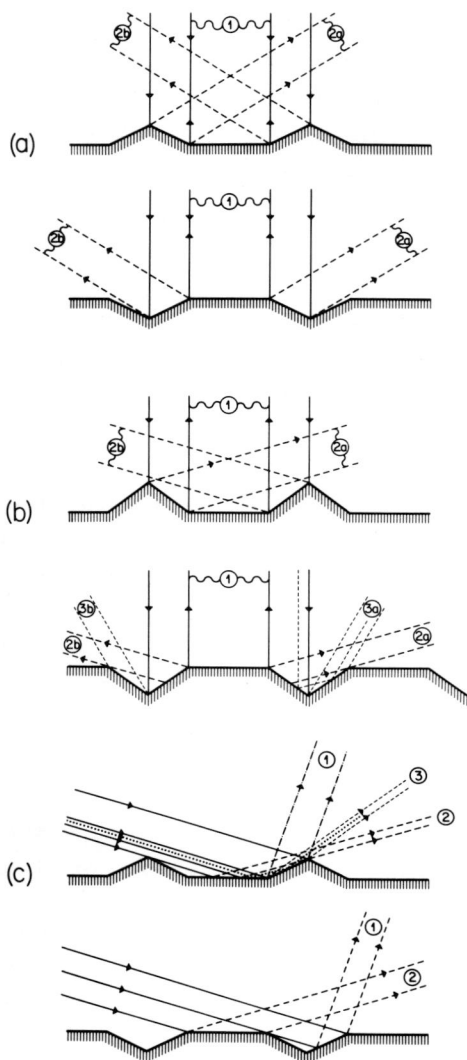

(a)

(b)

(c)

Fig.3.4a-c. Examples using classical particle scattering to demonstrate the influence of multiple scattering. In the case (a) of a shallow corrugation and normal incidence ($\theta_i = 0$) where no multiple scattering occurs, it is impossible to distinguish between $+\zeta(x)$ and $-\zeta(x)$, as both profiles give rise to the same intensity distribution of scattered particles; the quantum-mechanical analogue of this "single-hit" situation corresponds to the range of validity of the eikonal approximation. In cases (b) and (c), for grazing incidence and large corrugation amplitudes, $+\zeta(x)$ and $-\zeta(x)$ yield different intensity distributions. In quantum-mechanical diffraction, for such cases either the general method (Sect.3.3) or the GR method (Sect.3.4.1) or the iterative series (Sect.3.5) have to be used for intensity calculations

tinct scattering directions 1, 2a, and 2b are observed for both $+\zeta(x)$ and $-\zeta(x)$. In Fig.3.4b, the maximum corrugation amplitude is enhanced, and only for the case of $+\zeta(x)$ do we observe three scattering directions as in Fig.3.4c. For $-\zeta(x)$, however, due to double scattering in the trough, two further scattering directions 3a and 3b appear. A similar effect happens for the shallow corrugation of Fig.3.4a if we approach grazing angles of incidence (Fig.3.4c). In this case, the corrugation $+\zeta(x)$ gives rise to beams scattered in three different directions, whereas $-\zeta(x)$ gives only two scattering directions. For quantum-mechanical diffraction, the general conclusions are the same: for sufficiently small corrugations, no differences in the

Bragg peak intensities will occur for near-normal incidence for $+\zeta(\underline{R})$ and $-\zeta(\underline{R})$. This constitutes the region of validity of the eikonal approximation. For small corrugation amplitudes looked at by using grazing angles of incidence, as well as for large corrugations for any θ_i, differences in the Bragg intensities between $+\zeta(\underline{R})$ and $-\zeta(\underline{R})$ will occur. In the latter cases, for the calculation of intensities either the general method (Sect.3.3), the GR method (Sect.3.4.1) or the iterative series method of Sect.3.5 must be used. It should be emphasized, however, that in practice in many cases for grazing incidence and small corrugations as well as for large corrugations for any θ_i, the attractive part of the potential is no longer negligible and a reliable intensity determination is very difficult because of resonant scattering involving bound states of the potential (Chap.5).

To check the reliability of the results obtained by using the eikonal approximation, the unitarity condition (3.17) has to be fulfilled within at least a few percent. GARIBALDI et al. /3.2/ have developed kinematic factors which help to satisfy unitarity. We cite here an expression for the scattering amplitudes, which seems to be generally accepted to give the best results for intensity calculations within the eikonal approximation:

$$A_{\underline{G}}' = \frac{\underline{k}_i(\underline{k}_i-\underline{k}_{\underline{G}})}{k_{\underline{G}z}(-k_{iz}+k_{\underline{G}z})}\; A_{\underline{G}}^0 \quad , \tag{3.40a}$$

or explicitly expressed as a function of the angles $\theta_{\underline{G}}$ and $\phi_{\underline{G}}$, (3.13):

$$A_{\underline{G}}' = \frac{1+\cos\theta_i\cos\theta_{\underline{G}}\cos\phi_{\underline{G}}-\sin\theta_i\sin\theta_{\underline{G}}\cos\phi_{\underline{G}}}{\cos\theta_{\underline{G}}\cos\phi_{\underline{G}}(\cos\theta_i+\cos\theta_{\underline{G}}\cos\phi_{\underline{G}})}\; A_{\underline{G}}^0 \quad . \tag{3.40b}$$

We close this section with the remark that the frequently quoted Kirchhoff approximation is obtained by replacing $k_{\underline{G}z}$ by $-k_{iz}$ in (3.35).

3.5 Calculation of Intensities — Iterative Series

In this section, we outline a method developed by LOPEZ, YNDURAIN, and GARCIA /3.23/ which allows diffraction intensities to be calculated correctly for any scattering conditions $(\lambda_i,\theta_i,\gamma)$. The limits of convergence of this method have been shown for several numerically studied examples to be beyond that of the Rayleigh approach, and for small corrugations the method has the further advantage of fast convergence. Here, we follow a derivation provided by SCHLUP /3.24/ which starts from the hard-wall equation (3.28) in the Rayleigh limit

$$\sum_G A_G \exp[ik_{Gz}\alpha\zeta(\underline{R})] \exp[i\underline{GR}] = - \exp[ik_{iz}\alpha\zeta(\underline{R})] \quad . \tag{3.41}$$

Expanding the A_G's into a series using the strength of the corrugation α as an expansion parameter,

$$A_{\underline{G}} = \sum_{n=0}^{\infty} \alpha^n A_{\underline{G}}^{(n)} \quad , \tag{3.42}$$

and writing also $\exp[ik_{Gz}\alpha\zeta(\underline{R})]$ and $\exp[ik_{iz}\alpha\zeta(\underline{R})]$ in their Taylor series representation, (3.41) becomes

$$\sum_{\underline{G}} \left(A_{\underline{G}}^{(0)} + \alpha A_{\underline{G}}^{(1)} + \alpha^2 A_{\underline{G}}^{(2)} + \dots \right) \left(1 + \alpha i k_{Gz}\zeta(\underline{R}) + \alpha^2 \frac{(ik_{Gz})^2}{2!} \zeta^2(\underline{R}) + \dots \right) \exp[i\underline{GR}]$$

$$= - 1 - \alpha i k_{iz}\zeta(\underline{R}) - \alpha^2 \frac{(ik_{iz})^2}{2!} \zeta^2(\underline{R}) - \dots \quad . \tag{3.43}$$

Equating terms of equal order in α, we obtain for $n = 0$

$$\sum_{\underline{G}} A_{\underline{G}}^{(0)} \exp[i\underline{GR}] = - 1 \tag{3.44}$$

which yields

$$A_{\underline{G}}^{(0)} = - \delta_{G,0} \quad . \tag{3.45}$$

Equation (3.45) reflects the fact that for a completely flat surface [$\alpha = 0$ in (3.41)] the total intensity goes into the specular beam $\underline{G} = 0$. For $n = 1$, we obtain

$$\sum_{\underline{G}} \left[A_{\underline{G}}^{(1)} + A_{\underline{G}}^{(0)} ik_{Gz}\zeta(\underline{R}) \right] \exp[i\underline{GR}] = - ik_{iz}\zeta(\underline{R}) \quad . \tag{3.46}$$

Insertion of (3.45) into (3.46) yields for the second term of the left side
$\sum_{\underline{G}} A_{\underline{G}}^{(0)} ik_{Gz}\zeta(R) \exp[i\underline{GR}] = - ik_{Gz}\zeta(R) = ik_{iz}\zeta(R)$, so that $\sum_{\underline{G}} A_{\underline{G}}^{(1)} \exp[i\underline{GR}] = - 2ik_{iz}\zeta(R)$
or

$$A_{\underline{G}}^{(1)} = - 2ik_{iz}\zeta_{\underline{G}} \quad , \tag{3.47}$$

where $\zeta_{\underline{G}}$ denotes the Fourier transform of $\zeta(\underline{R})$ with respect to $\underline{G}$,

$$\zeta_{\underline{G}} = \frac{1}{\mathscr{F}} \int_{\substack{\text{unit} \\ \text{cell}}} \zeta(\underline{R}) \exp[-i\underline{GR}] \, d\underline{R} \quad . \tag{3.48}$$

Equating terms corresponding to n = 2, we obtain

$$\sum_{\underline{G}} \left[A_{\underline{G}}^{(2)} + A_{\underline{G}}^{(1)} \ ik_{\underline{G}z} \zeta(\underline{R}) + A_{\underline{G}}^{(0)} \ \frac{(ik_{\underline{G}z})^2}{2!} \ \zeta^2(\underline{R}) \right] \exp[i\underline{GR}] = - \frac{(ik_{iz})^2}{2!} \ \zeta^2(\underline{R}) \quad . \ (3.49)$$

Using (3.45), the last term on the left-hand side of (3.49) yields $-(ik_{iz})^2\zeta^2(\underline{R})/2!$ which cancels out the right-hand side, so that

$$A_{\underline{G}}^{(2)} = - \sum_{\underline{G}'} A_{\underline{G}'}^{(1)} \ ik_{\underline{G}'z} \zeta_{\underline{G}-\underline{G}'} \quad . \tag{3.50}$$

Proceeding in the same way to n = 3, we obtain

$$A_{\underline{G}}^{(3)} = - \sum_{\underline{G}'} \left[A_{\underline{G}'}^{(2)} \ ik_{\underline{G}'z} \zeta_{\underline{G}-\underline{G}'} + A_{\underline{G}'}^{(1)} \ \frac{(ik_{\underline{G}'z})^2}{2!} \ (\zeta^2)_{\underline{G}-\underline{G}'} \right] - 2 \ \frac{(ik_{iz})^3}{3!} \ (\zeta^3)_{\underline{G}} \tag{3.51}$$

from which the expression of the recursion formula for general n can already be anticipated,

$$A_{\underline{G}}^{(n)} = - \sum_{\underline{G}'} \left[\sum_{p=1}^{n-1} A_{\underline{G}'}^{(n-p)} \ \frac{(ik_{\underline{G}'z})^p}{p!} \ (\zeta^p)_{\underline{G}-\underline{G}'} \right] - \left(1-(-1)^n\right) \frac{(ik_{iz})^n}{n!} \ (\zeta^n)_{\underline{G}} \quad . \tag{3.52}$$

Using (3.45,47,50-52), the $A_{\underline{G}}^{(n)}$ can be calculated rather quickly, and once they have been established, the same set can be used to calculate the intensities for any α with (3.42). This means that for a given shape of the corrugation function $\zeta(\underline{R})$, the diffraction intensities can be obtained for any amplitude by variation of α. It is worth emphasizing that for negative values of α, the intensities for $-\zeta(-\underline{R})$ are obtained.

LOPEZ et al. /3.23/ have given another derivation of the series of $A_{\underline{G}}$, (3.41). They start from the general scattering equation (3.18) and arrive at equivalent (but less transparent) expressions for the $A_{\underline{G}}^{(n)}$. For this reason, the numerical results obtained with this method can be expected to be correct beyond the limit of convergence of the Rayleigh approach. Indeed, for some of the examples considered in /3.23/, the limit of convergence is almost a factor of three beyond the Rayleigh radius. However, in such cases the calculation has to be extended to large n (up to 70), and then, the method loses the advantage of being fast.

Having preselected the order n to which the calculation of the $A_{\underline{G}}^{(n)}$ is pursued, the range in α for which the method gives convergent results can be checked by verifying the unitarity condition (3.17).

3.6 A Few Illustrative Examples

Figure 3.5 shows the intensity of several diffraction beams as a function of the corrugation amplitude ζ_m for the case of a one-dimensional sinusoidal corrugation (3.29a). We assume normal incidence, $\theta_i = 0$, and have chosen $\lambda_i = 0.57$ Å and a = 3.52 Å. It must be noted, however, that any combination of λ_i, a, and ζ_m corresponding to the same ratios λ_i/a and ζ_m/a leads to the same results. The calculations were performed using both the iterative method (Sect.3.5) and the eikonal approximation (3.35,40). A glance at the topmost curve, which shows the sum of all diffraction intensities and which should satisfy the unitarity condition (3.17), shows that the case of the eikonal calculation is satisfied within < 1% even at $\zeta_m = 0.7$ Å, which is 40% above the Rayleigh limit (3.29b). Up to $\zeta_m = 0.6$ Å, the exact calculation using the iterative method of Sect.3.5 gives results practically indistinguishable from those shown in Fig.3.5. This proves that for small angles of incidence, the eikonal approximation can give reliable results even for rather large corrugation amplitudes.

It can further be seen from Fig.3.5 that for a completely flat surface ($\zeta_m = 0$), the total intensity is scattered into the specular beam [compare Sect.3.5, (3.45)].

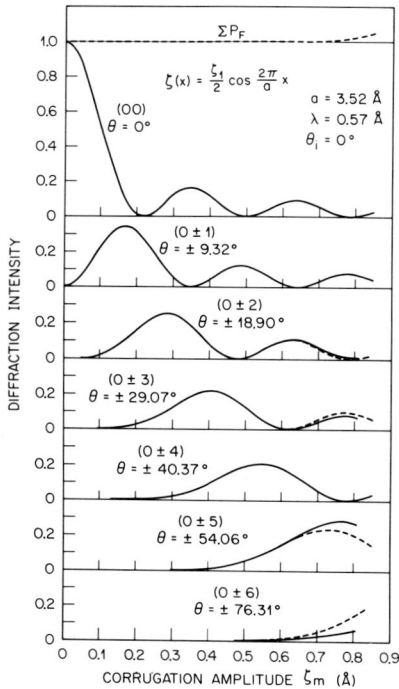

Fig.3.5. Intensity variation of several beams for a sinusoidal corrugation (3.29a) plotted as a function of the maximum corrugation amplitude ζ_m and calculated with the iterative series method (——) and the eikonal approximation (---). The particle beam impinges at normal incidence, $\theta_i = 0°$; the values of the lattice constant and the wavelength are assumed to be a = 3.52 Å and $\lambda_i = 0.57$ Å

For very small corrugation amplitudes ($\zeta_m = 0.07$ Å in Fig.3.5), aside the specular peak appreciable intensity is found only in the (0 ± 1) beams, and in this range their intensity is proportional to the square of the corrugation amplitude, $P_{(0 \pm 1)} \sim \zeta_m^2$. For further increasing corrugation amplitudes, the diffraction intensity is distributed in more and more beams, and the maximum intensity goes to diffraction peaks of higher and higher order. For instance, in Fig.3.5 for $\zeta_m = 0.35$ Å, the maximum intensity is observed for the (0 ± 3) beams. This result constitutes the quantum-mechanical analogue to the classical surface rainbow scattering illustrated in Fig.3.6 /1.7/. Following the classical trajectories of the particles for different impact locations b, it can be seen that the reflected intensity distribution is confined within the limiting angles θ_{RB} and θ_{RA}. The intensity scattered into a given angular range $d\theta$ is proportional to the corresponding range of impact parameters db, $P \sim db/d\theta$. The latter quantity tends to zero at the extreme angles θ_{RB} and θ_{RA}, so that the intensity has maxima at the edges of the distribution. The maxima can be traced to correspond to the steepest parts of the corrugation function. For a sinusoidal corrugation (3.29a) a simple geometrical derivation shows that the angle $\Delta\theta$ at which the rainbow angles occur as measured from the specular is given by

$$\Delta\theta = 2 \text{ arc tg} \frac{\pi}{a} \zeta_m \quad . \tag{3.53}$$

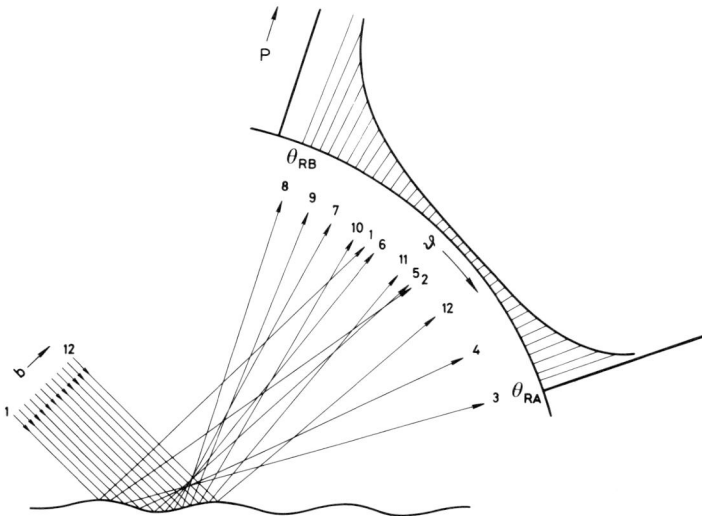

Fig.3.6. Classical surface rainbow scattering from a sinusoidal corrugation /1.7/

As observed from Fig.3.5, in the quantum regime appreciable intensity is found for beams outside the classical rainbow angles. If, therefore, (3.53) is used to calculate the maximum corrugation from the angular distance of the largest diffraction peak to the specular, its value is underestimated. Nevertheless, in many cases (3.53) can be used for a first rough estimate of the maximum corrugation amplitude. We finally note that the term "rainbow angle" stems from the optical rainbow observed in nature and is due to the presence of an extremum in the scattering of light rays from raindrops as a function of impact parameter showing up as a maximum of light scattered at this angle /3.2,25/.

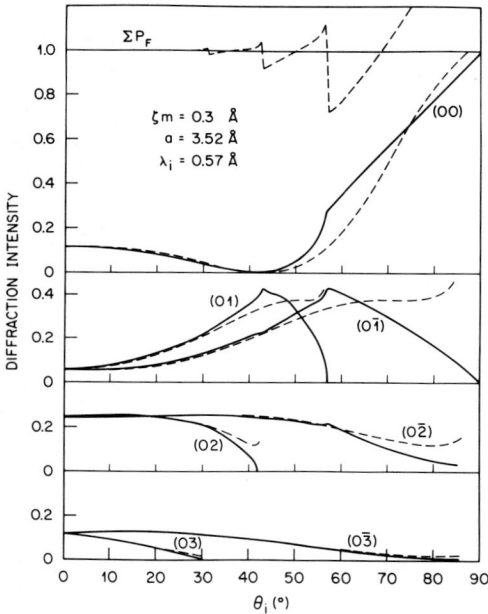

Fig.3.7. Intensity variation of several beams for a sinusoidal corrugation (3.29a) with $\zeta_m = 0.3$ Å plotted as a function of the angle of incidence θ_i and calculated by the iterative series method (——) and the eikonal approximation (---). The values of the lattice constant and the wavelength are the same as for Fig.3.5

Fig.3.7 shows the dependence of the intensity of several beams as a function of angle of incidence for the sinusoidal corrugation (3.29a) with $\zeta_m = 0.3$ Å and λ_i and a as in Fig.3.5. Results of calculations with both the eikonal approximation as well as the iterative series method (Sect.3.5) obtained in angular steps of 1° are exhibited. The topmost curve of Fig.3.7 shows that using the eikonal approximation, unitarity is satisfied within a fraction of a percent up to an angle of 30°, where a small but distinct structure in the course of $\sum_F P_F$ occurs. This is the region where with increasing θ_i the (03) peak reaches the horizon and vanishes. The deviations in the curve $\sum P_F$ come from the fact that the eikonal approximation fails to describe the threshold behaviour of the vanishing beams correctly [see the

dashed curves for (03), (02), and (01) in Fig.3.7]. Furthermore, the exact calculation using the iterative series shows that at angles where the beams (0n) vanish,
the intensities of the other beams exhibit cusp-like structures, which are not reproduced by the eikonal calculation. Nonetheless, for this case of a rather large
corrugation amplitude (85% of the Rayleigh limit), the eikonal approximation yields
satisfactory results up to θ_i of ~ 40°. For smaller corrugation amplitudes, the results for larger angles of incidence also can be trusted.

We mention at this point another shortcoming of the eikonal approximation which
is characteristic for two-dimensional corrugations. In such cases the specular intensity as a function of θ_i shows a different behaviour for different azimuthal
orientations of the sample relative to the incoming beam (angle γ in Fig.3.1). The
eikonal approximation is not able to reproduce this behaviour and gives an angular
dependence of the specular which is the same irrespective of the actual angle γ.
This can easily be seen from (3.35) as for $\underline{G} = (00)$ the expression for the scatter-
ing amplitude becomes $A_{(00)}^{(0)} = -(1/\mathscr{F}) \int \exp[2ik_{iz}\zeta(\underline{R})]d\underline{R}$, which does not contain any
quantity related to the angle γ (Fig.3.7).

Figure 3.8 shows the angular dependence of the intensities of a corrugation with
almost the same maximum amplitude as the one of Fig.3.7, but with a different shape
calculated with the iterative series method. The analytical expression for the corrugation function is

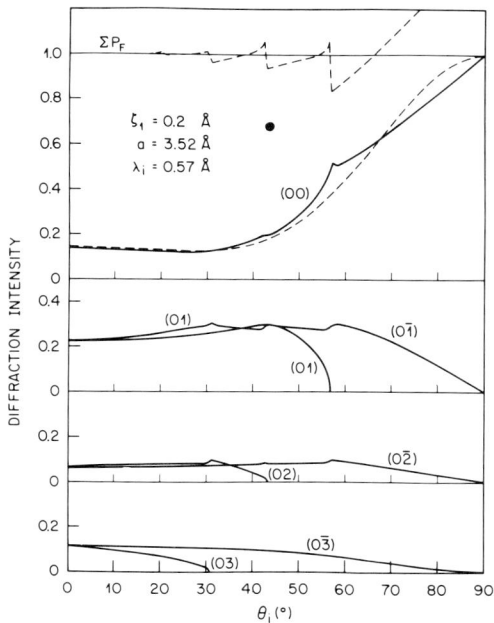

Fig.3.8. Intensity variation of several
beams for the corrugation function (3.54)
with $\zeta_1 = 0.2$ Å plotted as a function of
the angle of incidence θ_i and calculated
by the iterative series method. The dashed
lines for the total of all diffracted intensities $\sum P_F$ and for the specular correspond to results obtained with the
eikonal approximation. The values for
the lattice constant and the wavelength
are the same as for Fig.3.5

$$\zeta(x) = \frac{1}{2} \zeta_1 \left(\cos \frac{2\pi}{a} x + \frac{1}{2} \cos \frac{4\pi}{a} x \right) \quad . \tag{3.54}$$

The values for a, λ_i, and θ_i are the same for Fig.3.7; ζ_1 was chosen to be 0.2 Å so that the maximum corrugation amplitude ζ_m is 0.31 Å. Both the distribution of intensities at a fixed angle of incidence θ_i as well as the angular dependence of the intensities of all beams is quite different from that of Fig.3.7. This is important to notice as it shows that these differences are sensitive enough to determine the different forms of the corrugation functions from an analysis of the diffraction intensities. In the upper part of Fig.3.8, we again show results obtained with the eikonal approximation for both the sum of all diffracted beams as well as for the specular (dashed lines). In this case, even the vanishing of the (04) beam has a slight influence on the curve $\sum P_F$. Nevertheless, also here the eikonal approximation yields reasonable results for angles up to 30°.

3.7   The Inversion Problem

Up to now, we have dealt with the problem of obtaining diffraction intensities for a given corrugation function $\zeta(x,y)$ and a fixed scattering geometry $(\theta_i, \lambda_i, \gamma)$. In reality, however, one faces the inverse problem of determining the corrugation function $\zeta(x,y)$ from a set of measured intensity values. In most cases investigated so far, a Fourier series (3.2) with a limited number of coefficients was assumed, and by systematic variation of all coefficients those values were searched which gave the best agreement between the measured and calculated intensities for a given geometry. The degree of agreement can be judged by using a "reliability factor" which can be defined as

$$R = \frac{1}{N} \left[\sum_{\underline{G}} (P_{\underline{G}}^{calc} - P_{\underline{G}}^{exp})^2 \right]^{1/2} \quad , \tag{3.55a}$$

where N is the number of $\underline{G}$ vectors used for comparison between calculation and experiment; in this form, all the intensities are weighted equally. Another choice of the reliability factor, which weighs the different peaks according to their intensities, is

$$R_W = \frac{1}{N} \left[\sum_{\underline{G}} (P_{\underline{G}}^{calc} + P_{\underline{G}}^{exp})(P_{\underline{G}}^{calc} - P_{\underline{G}}^{exp})^2 \right]^{1/2} \quad . \tag{3.55b}$$

In searching for the best agreement between experimental and calculated intensities, it is usually worthwhile calculating both R and R_W and finding their minimum in parameter space.

The systematic search for the best-fit Fourier coefficients can be a rather tedious and time-consuming task, especially if an appreciable number of Fourier coefficients is needed to describe the scattering surface, since a rather dense mesh
in parameter space has to be investigated to find the minimum R values. The dense
mesh is necessary in order to find the real minimum, otherwise there is the possibility that one mistakes a local side minimum for the true one /3.26/. It should
be pointed out that due to experimental uncertainties, several combinations of
Fourier coefficients can often give the same R values, although the corresponding
corrugation functions look quite different. In such cases however, it is always
possible to determine the real corrugation function by investigating intensity sets
corresponding to different θ_i and λ_i.

CANTINI et al. /2.11/ made use of the Patterson series /3.27/ in their investigation of the corrugation function of NiO. This method has a definite advantage for
one-coefficient corrugations of the form (3.29a) or (3.30a) as the ζ_m values can be
directly determined using Bessel functions. For several Fourier coefficients, however, again a systematic variation of the coefficients has to be performed.

A very recent development /3.28/ allows a more direct approach of finding the
best-fit corrugation function. It rests on two observations. a) If both the abso-
lute values $|A_G|$ and the phases φ_G of the scattering amplitudes are known, the cor-
rugation function can be calculated to a rather high degree of accuracy from

$$\zeta(\underline{R}) = \frac{1}{2ik_i} \ln \left| - \sum_G A_G \exp(i\underline{G}\underline{R}) \right| \quad . \tag{3.56}$$

This equation is easily derived from (3.37) by approximating $\underline{k}_G$ by $-\underline{k}_i$. This approximation restricts the feasibility of the method to corrugation amplitudes $\zeta_m \sim 0.1a$.
b) In real situations, only intensities can be measured, and therefore, the phase
information is completely lost [compare (3.16)]. Hence, if only the absolute values
of the A_G's are known, it can be shown that approximate solutions of the hard-wall
scattering equation can be obtained with the phases of only a few intense diffraction beams approximately determined. This approximate determination is readily performed by investigating a coarse mesh of phases (usually steps of $\pi/2$ are sufficient).
In this way, approximate corrugations can be found and they can be used to generate
a full set of new phases φ_G which allows the calculation of an improved corrugation
function; this step can be repeated in a loop until optimum agreement between calculated and measured intensities is obtained. The effectiveness of the method was
proved for several one-dimensional model corrugations and successfully applied to
the case of H_2 diffraction from the quasi-one-dimensional corrugation of the adsorbate system Ni(110) + H(1×2) /3.26/.

3.8 Effects Due to the Softness of the Repulsive Potential

As discussed in Chap.2 and in Sect.3.1, the repulsive part of the particle-surface potential is in reality not infinitely steep. The influence of the softness of the repulsive potential on the diffraction intensities was quantitatively investigated by ARMAND and MANSON /3.29/. These authors succeeded in solving the scattering equation for both one- and two-dimensional corrugations using for the repulsive potential the exponential form

$$V(z) = C \exp\{-\kappa[z-\zeta(\underline{R})]\} \quad . \tag{3.57}$$

This special form of the potential was chosen because of its mathematical simplicity. Contrary to the hard-wall potential, the exponential potential allows wave penetration into the potential region to a degree dependent upon the κ values. For increasing κ, the wave penetration becomes less important, and for $\kappa = \infty$ the results correspond to the case of a hard corrugated wall. Model calculations of ARMAND and MANSON /3.29/ have shown that their numerical procedure converges only for very small corrugation amplitudes ($\zeta_m = 0.03a$). Nevertheless, from these calculations several important conclusions could be drawn. Firstly, the main effect of the finite slope of the potential is that scattering into the specular is enhanced at the cost of the other diffracted beams. This has the consequence that the application of the hard corrugated wall formalism in analysing intensity data from a soft potential tends to underestimate the corrugation amplitudes. Secondly, although the threshold behaviour of a beam vanishing below the horizon (Sect.3.3) is the same irrespective of the value of κ, the discontinuities observed at the threshold angles for the other beams (Figs.3.6,8) are affected differently for different κ /3.29/. ARMAND et al. have used the latter fact in an attempt to determine the value of κ for the case of H_2 diffraction from Cu(100) /3.30/. Figure 3.9 shows experimental intensities for the specular beam in the region where the (10) and (01) beams vanish below horizon. The wavelength of the H_2 beam used was 0.73 Å. Theoretical curves are shown in the same figure for the hard corrugated wall model ($\kappa = \infty$) and for an exponential potential with $\kappa = 6$ Å^{-1}. With the corrugation function in the form of (3.30a) and the best-fit value for $\zeta_m = 0.08$ Å (the lattice constant is a = 2.55 Å), this is the largest κ value for which the numerical procedure gave convergent results. The experimental values lie between the two theoretical curves and Armand et al. estimate the actual κ value to be about 8 Å^{-1}, which corresponds to a rather steep potential. In our opinion this result shows that even in the case of pure metal surfaces, where due to the free electrons the repulsive potential is expected to be soft, the application of the hard-corrugated-wall model in the analysis of

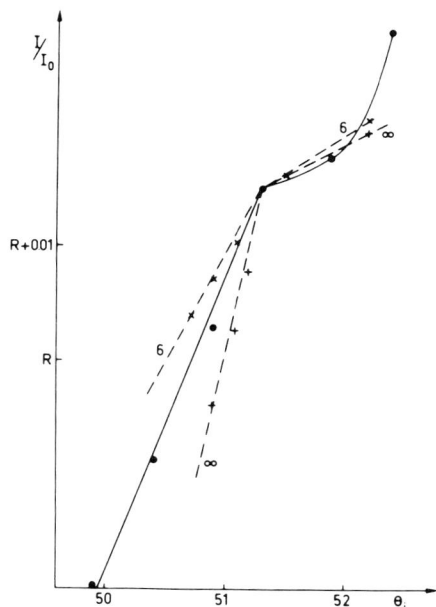

Fig.3.9. Variation of the intensity of the
specular beam of H_2 diffraction from Cu(100)
near the threshold of the (01) and (10) beams
(dots and full line). The dashed lines refer
to calculations using the hard-corrugated-
wall model ($\kappa = \infty$) and an exponential potential with $\kappa = 6$ $\text{\AA}^{-1}$ /3.30/

diffraction data will introduce only small errors in the determination of the form
and amplitude of the corrugation. Moreover, the influence of the softness of the
potential becomes less important, the larger the corrugation amplitudes. We close
this section with the remark that in view of Figs.2.3,7 the assumption of a single
κ is also a simplification as the steepness of the potential is somewhat different
for any point $\underline{R}$ within the unit cell and therefore in reality is described by a
function $\kappa(\underline{R})$.

4. Inelastic Scattering of Atoms from Surfaces

4.1 The Dependence of the Scattering on the Time Scale of the Interaction

The theoretical treatment for the calculation of diffraction intensities presented
in Chap.3 assumed that the atoms of the solid were at rest. However, both zero-point
motion and thermal vibrations of the surface atoms lead to inelastic scattering of the
incoming atoms. The influence of inelastic scattering on the intensities of x-ray
diffraction peaks was initially studied by DEBYE /4.1/ and WALLER /4.2/ and has been
reviewed by JAMES /4.3/. The principal effect of the thermal motion is a reduction

of the diffraction intensities without a change in the peak shape. The Debye-Waller factor, with which the diffraction intensity expressions in Chap.2 must be multiplied, can be written as

$$F_{DW} = \exp(-2W) \quad ,$$
(4.1)

where

$$W = \frac{1}{2} <(\underline{\rho} \cdot \underline{\Delta k})^2> \quad ;$$
(4.2)

$\Delta k = \underline{k}_i - \underline{k}_G$ is the momentum transfer in the scattering event, $\underline{\rho}$ is the displacement of a lattice atom from its equilibrium position and $<...>$ denotes thermal averaging. Note that for a given relative orientation of the incoming beam of the surface, W will be different for each diffraction peak.

The assumptions underlying (4.1,2) are that the interaction in the scattering event is both weak and short in duration. This can be justified for x-ray scattering and also for neutron scattering because of the short-range potential which leads to short-duration interaction despite the thermal neutron velocities. However, the application of the standard Debye-Waller treatment to atom scattering from surfaces is not straightforward, since the atom-surface interaction is strong and long-ranged and the incoming velocities are thermal. Therefore, the correction factor analogous to that of (4.1) for atom scattering will depend on the mass and velocity of the incoming atom, the time scale of the scattering interaction and the vibrational spectrum of the surface.

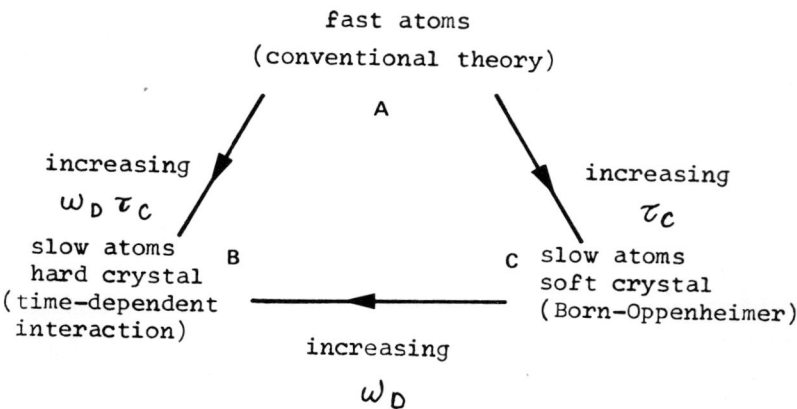

Fig.4.1. A graphical description of the different regimes of quantum scattering of atoms from surfaces /4.4/

This is illustrated in Fig.4.1, in which three limits to the scattering interaction are shown. For fast atoms, a result identical to that for x-rays and neu-
trons can be derived /4.4/. In this regime, which corresponds to vertex A in Fig.
4.1, the scattering atom is fast enough that the surface motion during the interaction can be neglected. The incoming atom then sees the disorder present at a
given time. For increasing interaction time τ_c, two different limits are observed
depending on whether the crystal has a hard or soft phonon structure. A measure of
the interaction time τ_c is given by $(\alpha v_i)^{-1}$, where α is a reciprocal range parameter for the potential, and v_i is the incoming-atom velocity; the hardness of the
crystal surface increases with its Debye frequency ω_D. For slow heavy atoms interacting with a soft crystal, the solid has time to adjust to the presence of the incoming atom, and the situation is similar to the Born-Oppenheimer approximation
with the surface atoms and the scattering particles taking the roles of the electrons and nuclei, respectively. This limit is shown as vertex C in Fig.4.1. For
slow atoms interacting with a hard crystal, the recoil of the surface atom can be
neglected, and the predominant effect of the increased interaction time is to aver-
age-out some of the surface disorder. This leads to an enhancement of the diffraction intensity when compared to fast-atom scattering and is shown as vertex B of
Fig.4.1.

This limit, which LEVI and SUHL /4.4/ call the time-dependent interaction regime, is the relevant limit for He, H_2, or Ne scattering from most surfaces. These
authors have dealt with this case using a semiclassical theory assuming a Morse
potential for the atom-surface interaction. The most important consequence of these
calculations, which will be discussed below, is that W in (4.2) cannot be separated
into separate momentum change and vibrational amplitude functions as can be done
for the fast-atom case. Due to the long-range attractive force (Chap.2) the particle will be accelerated along its trajectory, and in order to calculate the Debye-
Waller factor, the time-dependent atom-surface force and the time-dependent surface-
atom displacement must be calculated along the trajectory of the incoming atom.
These two quantities cannot be decoupled as for the fast-atom case.

4.2 The Debye-Waller Factor in the Time-Dependent Interaction Regime

For a solid with a Debye spectrum of vibrational frequencies, LEVI and SUHL /4.4/
obtain the result

$$W = \frac{12mE_{iz}kT}{\hbar M\omega_D^2} C \quad , \tag{4.3}$$

where C is a function of p and K, whereby $p = (1/2) \pi\omega_D\tau_c$ and $K = E_{iz}/D$. In (4.3)
M and m are the masses of a surface and gas atom, respectively. In these expressions
$E_{iz} = (1/2) mv_{iz}{}^2$, where v_{iz} is the perpendicular component of the incoming-atom
velocity at infinity, and D is the well depth of the Morse potential. C is shown as
a function of p for several values of K in Fig.4.2. It is seen that the presence of
an attractive potential will lead to a decrease of the diffraction intensities ($C > 1$)
for small p values which correspond to short interaction times. This consequence of
the attractive potential was initially reported by BEEBY /4.5/ and experimentally
verified by HOINKES et al. /4.6/. BEEBY proposed that the effect of the long-range
attractive potential on the diffraction intensities would be averaged out, since it
is due to an interaction with many surface atoms over a sufficiently long time. The
remaining effect of the attractive potential in Beeby's treatment is to accelerate
the incoming particle such that E_{iz} is increased by D. This model always predicts
a reduction in the diffracted intensities for $D > 0$.

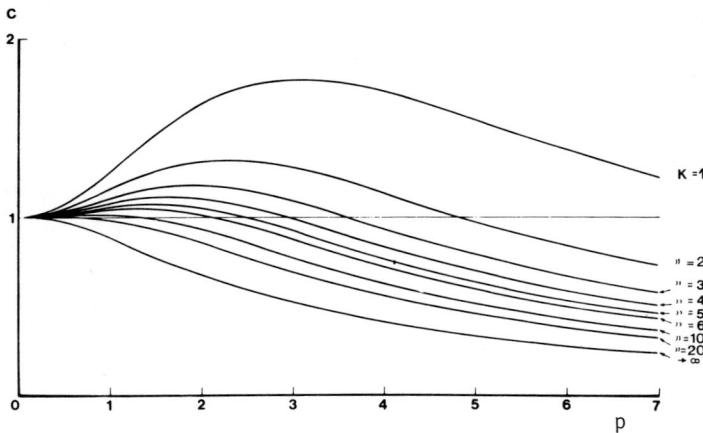

Fig.4.2. The dependence of the correction factor to the Debye-Waller exponent, C,
as a function of p, for several values of K /4.4/

However, the inclusion of thermal motion in the attractive potential in Levi and
Suhl's calculation shows that diffraction intensities can also be enhanced ($C < 1$)
by the presence of an attractive potential for large p corresponding to long interaction times as is seen in Fig.4.2. In practice, this requires both a collision time
longer than the Debye period and an energy which is high compared to the well depth.
In these calculations, the correction for the perpendicular energy of the scattering particle depends on p and K, but is always less than $E_{iz} + D$.

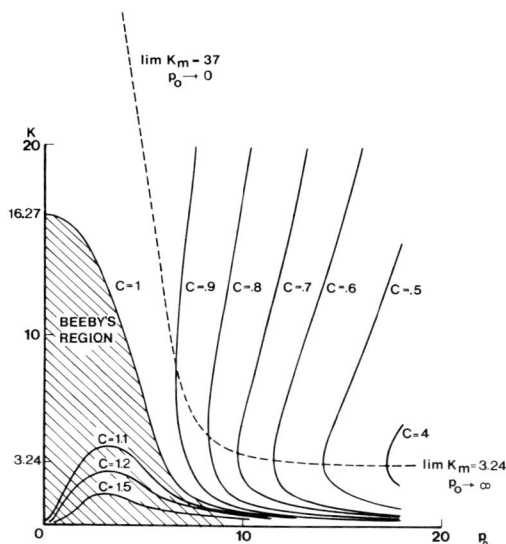

<u>Fig.4.3.</u> Constant C lines drawn in the (p_0,K) plane. The dashed line connects the K values yielding a minimum value of C for each p_0. The area in the lower left for which $C > 1$ is Beeby's region for which the diffraction intensities are reduced /4.4/

The effective correction factor C of Levi and Suhl is shown in Fig.4.3 as a function of p_0 and K, where $p_0 = p K^{1/2}$. This redefinition has the advantage that p_0 is independent of the energy of the incoming atom. The shaded area shows Beeby's region for which $C > 1$. If p_0 and K are known, and the use of a Morse interaction potential and a Debye spectrum for the solid are justified, C can be determined from Fig.4.3. Typical values for several gas-surface systems are shown in Table 4.1. Beam velocities characteristic of nozzle sources have been used, and α has been set equal to 2 $\overset{o}{A}^{-1}$. The last column of Table 4.1 shows values of C which have been obtained from Fig.4.3. With the exception of argon scattering from graphite, C is less than unity, showing that in general, the Beeby correction overestimates the influence of the attractive potential on the Debye-Waller factor. C is smallest for heavy atoms so that neon diffraction will be enhanced when compared with helium diffraction.

4.3 The Size Effect in the Debye-Waller Factor for Atom Scattering

The treatment discussed in Sect.4.2 [in particular, see (4.3)] has assumed that the scattering atom interacts with a single atom of the solid. However, the atom-surface potential is long-ranged so that this assumption is certainly not valid. Physically, this means that short wavelength oscillations in the surface will not be felt by the incoming atom. This leads to a reduction in W and to an enhancement of

Table 4.1. p_0, K, and p values for a number of gas-surface scattering systems. The velocity distribution assumed is characteristic of a nozzle beam, and the potential-range parameter is approximately 2 Å^{-1}. Also included in the last column are the C values taken from Fig.4.3 /4.4/

	p_0	K	p	C
He/LiF	0.12	9	4	0.6
Ne/LiF	0.19	4	9	0.4
Ar/LiF	0.16	1.6	12	0.5
He/Graphite	7.5	4.5	3.5	0.9
Ne/Graphite	11	2	8	0.7
Ar/Graphite	9.5	0.8	10.5	1.2

the diffraction intensities. A correction for this effect was originally proposed by HOINKES et al. /4.6/ who used a model in which the motion of the surface atoms was uncorrelated. ARMAND et al. /4.7,8/ have refined this idea to include the correlated motion of surface atoms on the (100) plane of face-centred-cubic (fcc) crystals. In this model, only harmonic forces between nearest neighbours are included and the force constant is adjusted to give the correct value of the bulk Debye temperature. Assuming that the incoming atom interacts with the four atoms of the surface unit cell of Cu(100), the normal component of $\langle\rho^2\rangle$ can be calculated for this configuration /4.9/. It is found that the correlated mean-square displacement is only 40% of that for an individual surface atom. Of course, the number of surface atoms with which the incoming atom interacts depends on the range of the interaction so that at the present state of knowledge of atom-surface potentials, it is not easy to correct for this effect unambiguously. However, since the mean-square correlated displacement will be smaller, the greater the number of atoms considered, the enhancement of the diffraction intensities is greater, the longer the range of the interaction potential.

4.4 Experimental Investigations of the Debye-Waller Factor for Atom-Surface Scattering

Experimentally, W can be determined by measuring the dependence of a diffracted beam intensity on either the surface temperature T_s or on the normal component of the incoming-atom energy E_{iz}. For the latter case, either the total energy of the atom or the angle of incidence can be varied, since $E_{iz} = E_i \cos^2\theta_i$. The determination of W by varying the angle of incidence is only possible for very small corrugations for

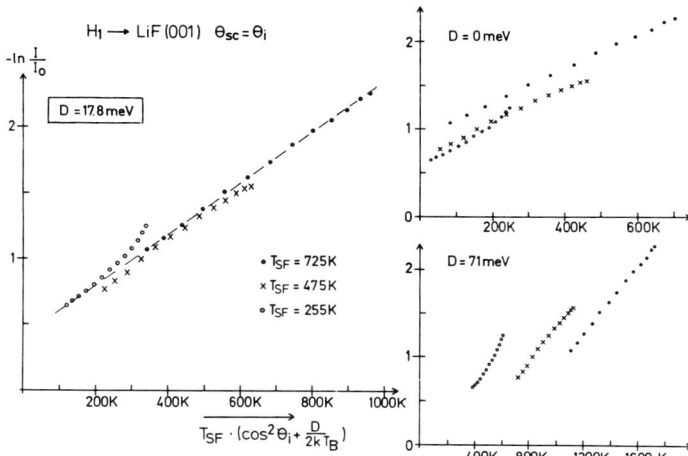

Fig.4.4. Logarithmic plot of the specular intensity for H scattering from LiF(100) versus the product of surface temperature and momentum transfer perpendicular to the surface. The left side shows a fit for an attractive potential D = 17.8 meV. The right side shows plots for D = 0 and 71 meV [Ref. /4.6/, the corrected version of Fig.5]

which only the specular peak is intense, since otherwise the angular variation can be dominated by structural effects. Figure 4.4 shows the results of a measurement of the specular-beam intensity for H diffraction from LiF(100) in which the angle of incidence has been varied for three surface temperatures and a Beeby correction has been made /4.6/. The attractive well depth is known to be 17.8 meV from resonant scattering measurements /4.10/. For this value of D, it is seen that all points lie near the straight line for the Beeby correction, but that substantial deviations are observed for D = 0 and 71 meV. Using a model of uncorrelated surface-atom displacements, the surface Debye temperature is found to be 415 ± 42 K. This is rather small when compared with the bulk value of 732 K and may be due to the assumptions of uncorrelated motion and that C = 1.

Detailed studies of the Debye-Waller factor for He and Ne scattering from Cu(100) have been carried out by LAPUJOULADE et al. /4.11/. Figure 4.5 shows an angular scan for Ne scattering at two substrate temperatures. It is seen that the inelastically scattered intensity has a broad lobular distribution whose center is shifted from the specular direction towards the surface normal. Whereas the elastic scattering can be clearly seen in Fig.4.5 for Cu(100), this is not the case for all metal surfaces. Figure 4.6 shows angular distributions taken in the authors' laboratory for Ne scattering from the (1×2) reconstructed surface of Au(110) for which He diffraction traces are shown in Figs.6.7 and 9.4. A double lobe which is far removed from the specular angle is seen for in-plane scattering with the beam incident in the [001] azimuth, and a single lobe centered near the specular angle is seen for the

I/I_0

0.0050

I   T = 473K

0.0025

0
0.075   55   65   75   85   90 θ_r

II   T = 70 K

0.050

0.025

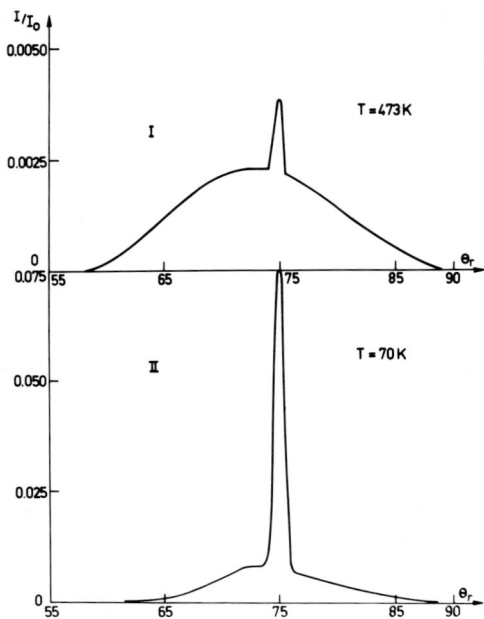

Fig.4.5. Specular-beam intensity as a function of the scattering angle for Ne scattered from Cu(100) for surface temperatures of 70 K and 473 K. E_i = 63 meV, θ_i = 75° /4.11/

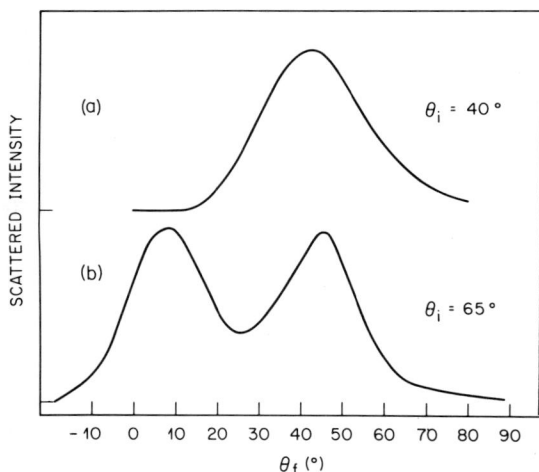

0
55   65   75   85   90 θ_r

SCATTERED INTENSITY

(a) θ_i = 40°

(b)

θ_i = 65°

-10   0   10   20   30   40   50   60   70   80   90
θ_f (°)

Fig.4.6a,b. In-plane scattered intensity as a function of the scattering angle for Ne scattering from the (1×2) reconstructed Au(110) surface in (a) the [1$\bar{1}$0] and (b) the [100] azimuths. The surface and nozzle temperatures are 300 K

beam incident in the [1$\bar{1}$0] azimuth. No elastically scattered intensity was observed for surface temperatures as low as 100 K and for a wide range of beam energies and angles of incidence. This shows that the Debye-Waller factor is appreciably smaller for Au(110) than for Cu(100).

Figures 4.7,8 show experimental results of LAPUJOULADE et al. /4.11/ for the dependence of the specular-beam intensity on the surface temperature for Ne and He scattering from Cu(100). Also shown in these figures are theoretical fits to the data using the fast-atom limit for W,

$$W = \frac{1}{2} (\Delta k)^2 \langle \rho_z^2 \rangle \quad , \qquad (4.4)$$

where only the vibrations perpendicular to the surface have been taken into account. The mean-square correlated displacement $\langle \rho_z^2 \rangle$ has been calculated as outlined in Sect. 4.3 and multiplied by a correction factor λ which was used as a fit parameter. The attractive well depth was introduced using the Beeby correction. The best-fit parameters giving the solid lines in Figs.4.7,8 are $\varepsilon = 9.8$ meV for He and 124 meV for

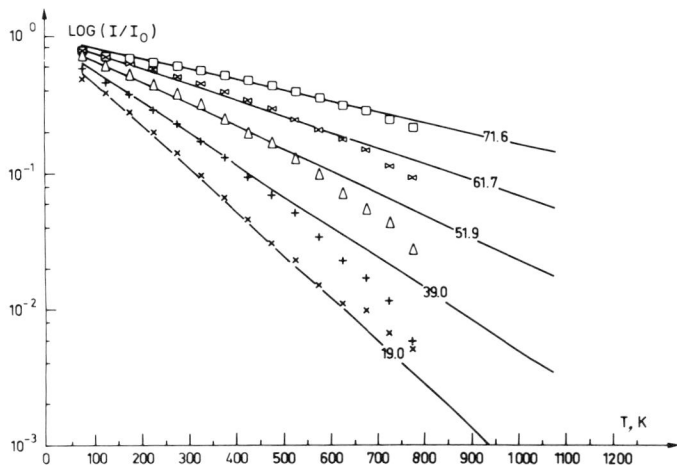

Fig.4.7

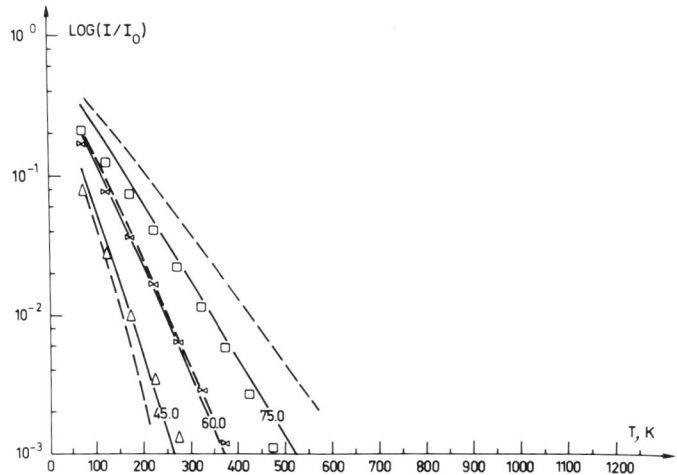

Fig.4.8

Fig.4.7. Specular-beam intensity versus the surface temperature T for He scattering from Cu(100). The parameters are the angles of incidence and $E_i = 63$ meV. The solid lines are a theoretical fit to the data /4.11/

Fig.4.8. Specular-beam intensity versus the surface temperature T for Ne scattering from Cu(100). The parameters are the angles of incidence and $E_i = 63$ meV. The solid and dashed lines are theoretical fits to the data /4.11/

Ne and $\lambda = 0.855$ for He and 0.775 for Ne. The authors discuss several interpretations for the deviation from $\lambda = 1$, which would be expected if the incoming atom
interacted with four Cu atoms coupled by harmonic forces, and the force constant
was determined as outlined in Sect.4.3. One possibility is that the force constant
between the surface atoms is higher than assumed. However, the deviation of λ from
unity may also be due to the range of the potential, and $\lambda > 1$ indicates that the
incoming atom interacts with more than four Cu atoms. The well depth deduced for He
is within the range of what is expected for metal surfaces, but that for Ne is much
higher than would be expected. However, the assumption of $D = 60$ meV gives a poor-
quality fit as shown by the dashed line in Fig.4.8.

These measurements have been reinterpreted by BÜHEIM et al. /4.12/. They point
out that besides the finite-time and finite-size effects outlined in Sects.4.2,3,
there are further contributions to the Debye-Waller exponent resulting from inelastic processes with momentum transfer parallel to the surface. This can lead to an
increase of the exponent by factors of the order of two to three, thus compensating
the finite-size effects to a large extent. BÜHEIM et al. consider it best to neglect
both corrections and to compare the fit parameters of the potential with additional
information on inelastic processes such as energy accommodation coefficients using
a theory /4.13/ consistent with the approximations for the Debye-Waller factor.
Their fit to the neon data of Fig.4.8 using a Morse potential with $D = 33$ meV is
shown in Fig.4.9. Note the curvature in the plot at low temperatures, which is due
to the inclusion of the zero-point motion of the solid. The value of D obtained with
this treatment is within the range expected for the Ne-metal surface potential /4.14/
and the good agreement obtained with experiments shows that the theoretical formulation of the Debye-Waller factor for atom scattering has advanced to the stage where
quantitative agreement can be achieved.

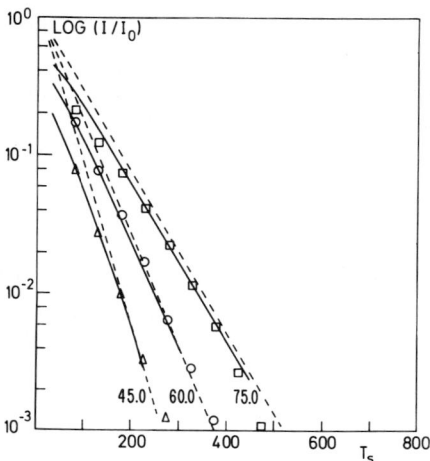

Fig.4.9. Specular-beam intensity versus the
surface temperature T for Ne scattering from
Cu(100). The data points and the dashed line
are from /4.11/. The solid lines are a fit
with $D = 33$ meV taken from /4.12/

In closing this section, we shall briefly mention a number of theoretical and experimental studies relevant to the Debye-Waller factor in atom-surface scattering. ASADA /4.15/ has published a detailed study of He and H_2 scattering from Ag(111). This paper also includes an extensive list of references to these investigations. MÜLLER-HARTMANN et al. /4.16/ have suggested that inelastic scattering through direct excitation of electrons near the Fermi level may be an important mechanism in metals. MASON and WILLIAMS /4.17/ have reported results for He scattering from Cu(110) in which the specular intensity shows an anomalous dependence on Δk. This result has been discussed elsewhere /4.11/. Recent results by CANTINI et al. /4.18/ show that resonant scattering can also couple with inelastic scattering.

5. Influence of the Attractive Part of the Potential on Diffraction Intensities

5.1 Modifications for the Calculation of Diffraction Intensities

In Chap.3, we presented an extensive discussion of how diffraction intensities can be calculated for a given corrugation and a given scattering geometry, neglecting the influence of the attractive part of the potential. The existence of the attractive well leads, however, to the important phenomenon of resonant scattering or selective adsorption, which can influence the Bragg intensities considerably. We will discuss this effect in the following sections. In this section, we consider simple modifications in the intensity calculations which must be performed in situations where the depth of the attractive part D cannot be neglected in comparison to the incoming energy E_i, but where selective adsorption does not play a role. For this purpose, it is sufficient to characterize the attractive well solely by its depth D /3.2,5.1/.

Due to their attractive interaction with the surface, the particles are accelerated in the region of the attractive well, and their effective energy perpendicular to the surface E'_{iz} is increased against E_{iz} according to

$$\frac{2m}{\hbar^2} E'_{iz} \equiv k'^2_{iz} = k^2_{iz} + \frac{2m}{\hbar^2} D \equiv (E_{iz} + D) \frac{2m}{\hbar^2} \quad . \tag{5.1}$$

This has the consequence that the particle beam is refracted toward the surface normal in the region of the attractive well, so that the beam hits the repulsive wall under a smaller effective angle of incidence θ_i' given by

$$\sin\theta_i' = \frac{k_i}{k_i'}\sin\theta_i \quad , \tag{5.2}$$

the total effective energy of the particles then being

$$E_i' \equiv \frac{\hbar^2}{2m}k_i'^2 = \frac{\hbar^2}{2m}(\underline{k}_{iz} + \underline{K})^2 + D \equiv E_i + D \quad . \tag{5.3}$$

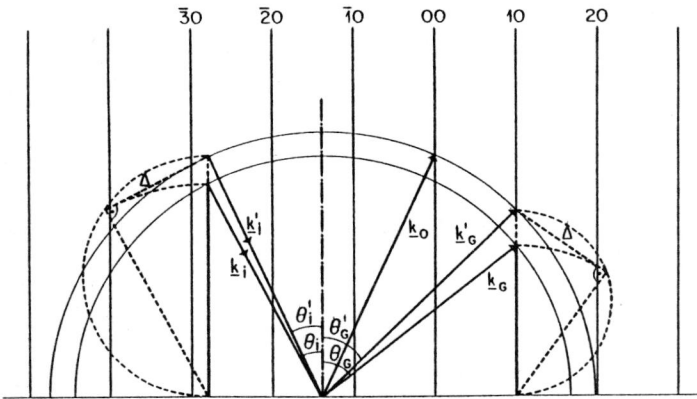

Fig.5.1. Ewald construction for diffraction of particles with wave vector k_i and angle of incidence θ_i. Due to the attractive potential of depth D, the particles are accelerated towards the surface according to (5.1). The incoming beam is therefore refracted towards the surface normal. The Ewald construction for the effective k_i' and θ_i' is also shown. The dotted lines represent a geometrical proof that the effective angles of emergence θ_G' correspond to the ones actually observed, θ_G, as the particles lose the extra energy D when leaving the surface region. [Compare (5.4); $\Delta = (2mD/\hbar^2)^{1/2}$]

The situation is shown in Fig.5.1. The solid lines refer to the Ewald construction for the initial wave vector $\underline{k}_i$, and to the Ewald construction for $\underline{k}_i'$. For the latter, the z-component of the diffracted beams $\underline{G}$ is given in analogy to (5.1) by

$$k_{Gz}'^2 = k_i - (\underline{K} + \underline{G})^2 + \frac{2mD}{\hbar^2} \quad . \tag{5.4}$$

The diffracted particles will lose their extra energy D when leaving the region of the attractive well and are therefore refracted from the surface normal, so that they finally occur at diffraction angles corresponding to $\underline{k}_i$. A geometrical proof that the effective diffraction angles θ_G' and the real angles θ_G are in accordance with (5.4) is given by the dotted construction shown in Fig.5.1. [Note that $\Delta = (2mD/\hbar^2)^{1/2}$].

Therefore, to a first approximation the attractive well can be taken into account by replacing k_i and θ_i by k_i' and θ_i' in the corresponding expressions of Chap.2 used for intensity calculations. This is certainly a valid procedure, as long as the Ewald spheres for k_i and k_i' cover the same range of $\underline{F}$ vectors as is the case in Fig.5.1. If the Ewald sphere for $\underline{k}_i'$ comprises more $\underline{F}$ vectors than that for $\underline{k}_i$, intensity will be calculated for these extra beams although in reality they cannot be observed ($\theta_F' < 90°$, $\theta_F > 90°$). Hence, this simple procedure will give reliable results only as long as the intensities of the extra beams come out very small. This will always be the case for small corrugations, if the beam comes in near normal incidence. Otherwise, the full mathematical apparatus outlined in Sect.5,3, which takes into account resonant scattering, has to be applied.

5.2 Bound Surface States and Resonant Transitions

Already in their early He-diffraction experiments, STERN and coworkers observed characteristic structures (usually minima) in the diffraction peaks as a function of angle of incidence θ_i /5.2/ or as a function of azimuthal angle γ (Fig.3.1) of the sample relative to the incoming beam /5.3/. In Fig.5.2, we show a recent example for the latter case obtained by FINZEL et al. /2.2/ using the diffraction of atomic deuterium from a NaF(100) surface. LENNARD-JONES and DEVONSHIRE /5.4/ attributed these features correctly to resonant transitions of the incoming particles into bound states of the attractive potential and referred to the phenomenon as "selective adsorption". The basic underlying idea is simply related to the Bragg condition for conservation of momentum parallel to the surface

$$\underline{K} + \underline{G} = \underline{K}_G \tag{3.8}$$

in connection with the condition for conservation of the total particle energy

$$k_i^2 = k_{\underline{G}}^2 = (\underline{K} + \underline{G}) + k_{Gz}^2 \quad . \tag{3.9}$$

The normal component of the energy of the particles

$$E_{Gz} \equiv \frac{\hbar^2}{2m} k_{Gz}^2 = \frac{\hbar^2}{2m} \left[k_i^2 - (\underline{K} + \underline{G})^2\right] \equiv E_i - E(\underline{K}_G) \tag{5.5}$$

is always ≥ 0 for particles scattered into Bragg peaks $\underline{F}$ [compare (3.11)]. The existence of the attractive part of the potential has the consequence that the particles can gain kinetic energy at the expense of potential energy if conditions (3.8,9) per-

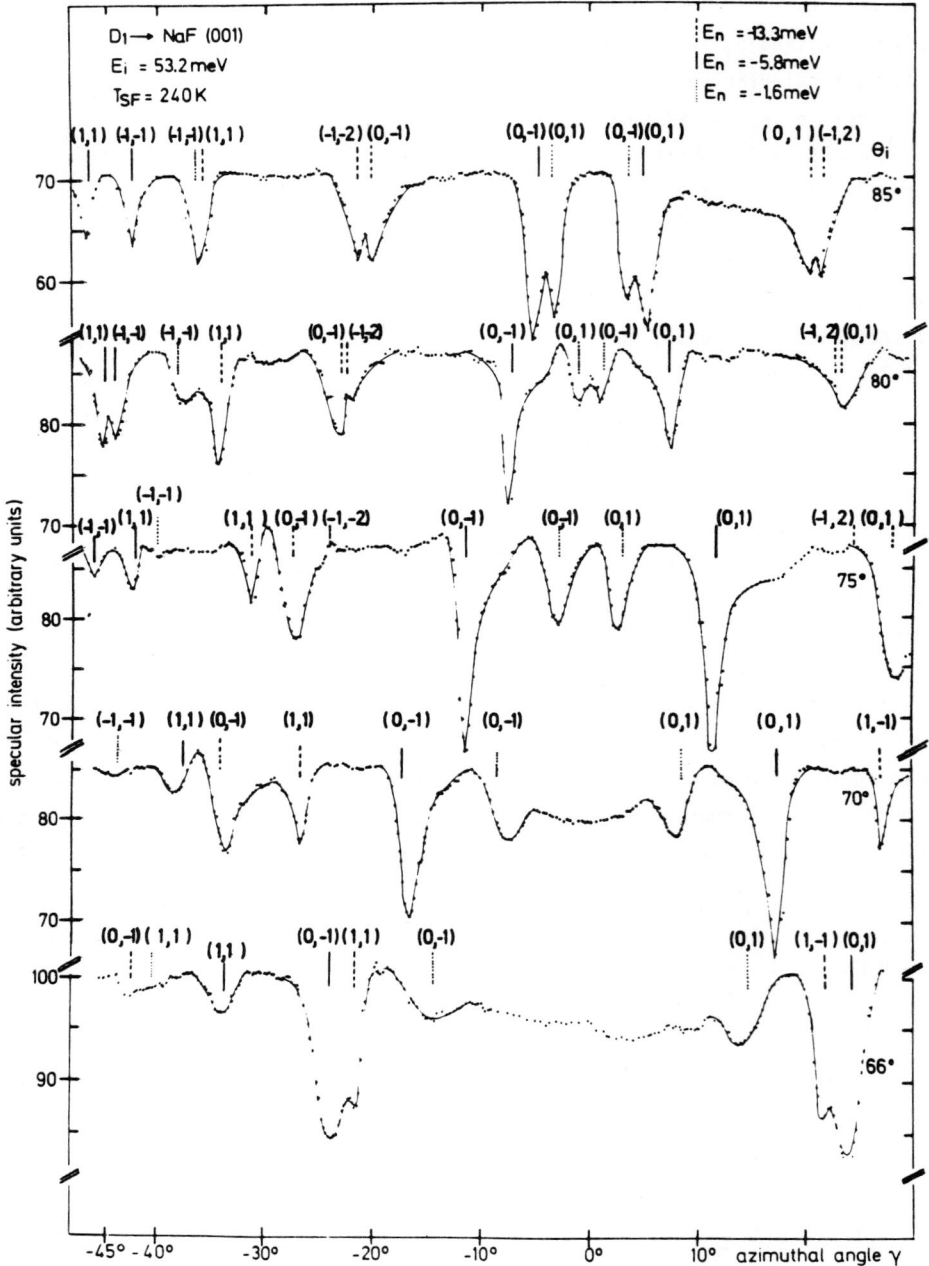

Fig.5.2. Resonant structures observed with atomic deuterium scattered from NaF(001). The dependence of the specular intensity on the azimuthal orientation γ of the sample relative to the incoming beam is shown for several angles of incidence θ_i and a fixed incident energy E_i. Transitions to bound states with the corresponding reciprocal lattice vectors and binding energies are indicated /2.2/

mit a transition into one of the allowed energy levels ε_ν of the attractive potential via a vector $\underline{G}$ outside of the Ewald sphere. In such a case, the particles move
parallel to the surface in the direction $\underline{K}_G$ with a total energy $\hbar^2 k_i^2/2m + |\varepsilon_\nu|$, but
are bound perpendicular to the surface with the binding energy

$$\varepsilon_\nu = E_{\underline{G}z} < 0 \quad , \tag{5.6}$$

where $E_{\underline{G}z}$ is determined by (5.5). Whenever such a transition to a bound state is
possible, the intensity of the observed Bragg peaks $\underline{F}$ will be modified, and the location of the resonant structures as a function of λ_i, θ_i, and γ can be used for
determination of the bound-state energies according to (5.5). With the intensity
data for the (00) beam shown in Fig.5.2, FINZEL et al. /2.2/ were able to determine
the three deepest energy levels of the system $D_1/\text{NaF}(100)$. The corresponding $\underline{G}$ vectors by which the resonant transitions are obtained are indicated in the figure.
Another energy level ε_3 could be traced by studying the angular dependence of the
(10) beam. All the bound-state energies listed in Table 2.1 were obtained from similar experiments.

Rearranging (5.5) to the form

$$(\underline{K}_i + \underline{G})^2 = (E_i + |E_{\underline{G}z}|) \frac{2m}{\hbar^2} \quad , \tag{5.7}$$

one can see that the $\underline{K}_G$ vectors leading to a particular transition $\varepsilon_\nu = E_{\underline{G}z}$ lie on
a circle centred at $\underline{G}$ with the radius $[2m(E_i + |E_{\underline{G}z}|)\hbar^2]^{1/2}$. Figure 5.3 depicts this
situation graphically, and Fig.5.4 shows an experimental result obtained by MEYERS
and FRANKL /5.5/ for He diffraction from NaF(100), proving that (5.8) is fulfilled
reasonably well. This implies that the bound particles may be treated as nearly free
in two dimensions parallel to the surface, i.e.

$$E(\underline{K}_G) = \frac{\hbar^2}{2m} K_G^2 \quad . \tag{5.8}$$

In Fig.5.3, we have plotted a particular vector $\underline{K}$ which leads to fulfillment of
(5.8) for both the $\underline{G}$ vectors $(0\bar{1})$ and $(\bar{1}\bar{2})$. In Fig.5.5a, we present the experimental results of HOINKES et al. /2.3/ for exactly the same situation using D_1 dif-
fraction from NaF(100). The arrows indicate the crossover of the resonance minima
for $\varepsilon_0 = -13.3$ meV as expected from (5.9) for the reciprocal lattice vectors $(0\bar{1})$
and $(\bar{1}\bar{2})$. In contrast to this expectation, two well-separated minima are always observed in the experiment. In accordance with the theoretical investigations of CHOW
and THOMPSON /5.6/, this effect could be explained by energy splitting of mixed de-
generate bound states. The two resonant bound states $|\nu(j\ell)\rangle = |0(0\bar{1})\rangle$ and $|0(\bar{1}\bar{2})\rangle$

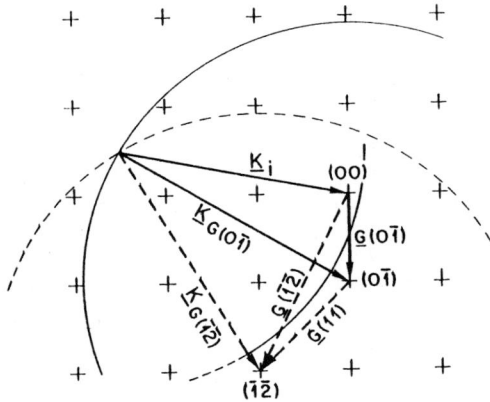

Fig.5.3. Graphical representation of (5.8) showing that the K_G vectors leading to a particular resonant transition (ϵ_ν, G) lie on a circle with radius $[(E_i + |\epsilon_\nu|)2m/\hbar^2]^{1/2}$ centred at G. For the case indicated, resonant transitions to both $G = (0\bar{1})$ and $G = (\bar{1}\bar{2})$ are possible. In such situations, band-splitting effects can be observed

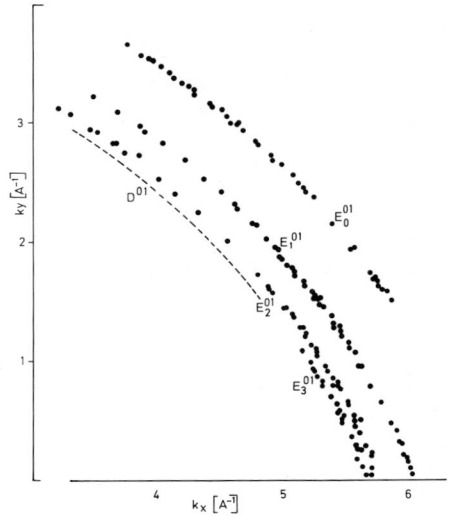

Fig.5.4. x- and y-components of the wave vector K for the (01) selective adsorption transitions ($\nu = 0$-3) for He/NaF(100), proving experimentally (5.8) /5.5/

belonging to the same total energy $E_{0(0\bar{1})} = E_{0(\bar{1}\bar{2})}$ are mixed near the resonance angle because of the influence of the periodic term $v_{(0\bar{1})-(\bar{1}\bar{2})} = v_{11}$ of the potential, (2.1). For the mixed states, perturbation theory yields the energy eigenvalues

$$E_{a,b} = \frac{1}{2}\left[E_{0(0\bar{1})} + E_{0(\bar{1}\bar{2})} \pm \sqrt{\left(E_{0(0\bar{1})} - E_{0(\bar{1}\bar{2})}\right)^2 + 4H_{12}^2}\right] \qquad (5.9)$$

with $H_{12} = b_{11}D\exp(\sigma^2\hbar/m\omega)$ [compare (2.13)] and $\omega = 2(D-|E_0|)\hbar$. Using (5.9), the only unknown parameter b_{11} could be calculated from the experimentally observed splitting. The results are summarized in Fig.5.5b. The circles indicate the experimentally observed locations of the minima, their darkness being roughly proportional to their intensity. The full heavy line and the dashed line correspond to the full and open arrows of Fig.5.5a, respectively. These lines cross in accordance with (5.8). The full light lines correspond to a calculation with (5.9) using the best-fit value for $b_{11} = 0.02$. The corresponding contribution to the potential is shown in Fig.5.2.

The observed energy splitting corresponds to a two-dimensional band-structure effect and is analogous to the energy splitting near zone boundaries of electrons

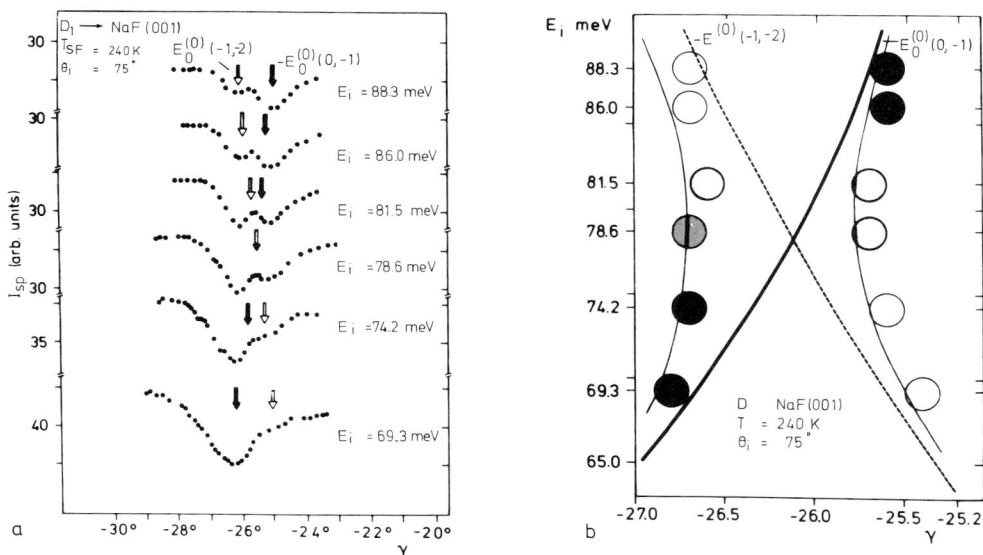

Fig.5.5. (a) Experimentally observed splitting of bound-state resonance minima in the specular intensity of atomic deuterium scattered from NaF(100) as a function of azimuthal orientation γ for different incoming energies E_i. (Compare Fig.5.4.) The contributing bound channels are indicated by arrows. (b) Calculated dependence of resonances on E_i and γ for independent bound channels (dashed and full heavy lines) and for mixed bound channels (full light lines). The circles indicate the experimentally observed locations of the resonances /2.3/

moving in periodic potentials. It shows that (5.8) is only approximately true; the particles moving in bound states parallel to the surface feel the two-dimensional periodicity of the surface potential and their wave functions are not plane waves as for free particles, but Bloch waves, whose energy spectrum contains gaps. Therefore, (5.8) fails to hold near intersections of the circles described by (5.7), (Fig.5.2). The most important consequence of this effect lies in the fact that the dominant higher-order terms in the potential series can be determined to a high accuracy. Similar band splitting has been observed for He/NaF(100) /5.7/, He/LiF(100) /5.8/ and He/graphite(1000) /2.8,9,37/.

It should be noted that in the intensity variation of the specular beam due to resonant scattering, mostly minima, but sometimes also maxima and Fano-type anti-resonances are observed. On the basis of a purely elastic theory, WOLFE and WEARE /5.9/ have worked out three rules which predict under which conditions which structures should appear. As these rules can be helpful in analyzing experimental data and may give a rough idea of which Fourier components of the potential are important, it is worth citing them here: a) Specular minima will be observed when the channel in resonance couples directly and strongly to the specular channel (more strongly than through indirect coupling) and to at least one other open channel. b) Mixed maxima-minima (maxima predominant) in the specular intensity will be observed when the only open channel to which the resonant state couples strongly and directly is the specular. c) Specular maxima will be observed for resonant channels that couple only indirectly to the specular through strong Fourier components. Examples for these rules are given in the original paper of WOLFE and WEARE /5.9/ as well as in /5.10/. It must, however, be emphasized that the "Weare-Wolfe" rules are derived on the basis of a purely elastic theory, and therefore, deviations can occur in cases where inelastic effects play an important role.

CHOW and THOMPSON /5.6/ have also considered theoretically the effect of strong coupling of bound channels to diffracted channels. An example for such a situation is shown in Fig.5.6 which refers to H-diffraction from KCl(100) covered with H_2O as

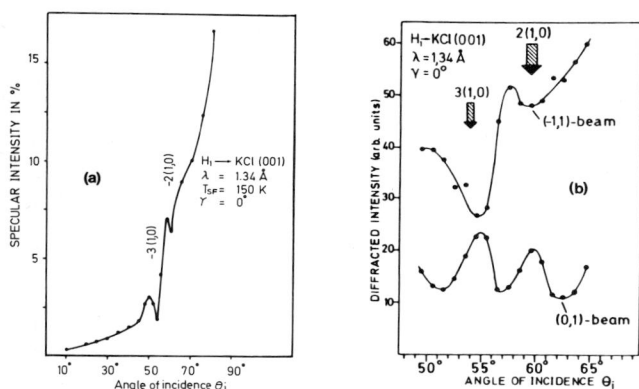

Fig.5.6. (a) Specular intensity as a function of angle of incidence θ_i for atomic hydrogen (E_i = 46 meV) scattered from KCl(001). Bound-state minima are indicated. (b) Diffracted intensities for ($\bar{1}1$) and (01) beams versus θ_i. At resonance the (01) beam shows pronounced maxima due to strong coupling of the resonant channel (10) to the open channels (0 ±1) via the strong potential term v_{11} /2.3/

investigated by HOINKES et al. /2.3/. In this case, intensity minima are found for the specular as well as for the ($1\bar{1}$) beam at rather large angles of incidence θ_i. These minima correspond to intensity losses via the channel (10) into the bound states ε_2 = -15.9 meV and ε_3 = -10.3 meV /2.4/. The (01) beam, however, shows intensity maxima at the same values θ_i. As is easily observed from geometrical considerations, the symmetric open diffraction channels (0 ±1) are those coupled strongly via v_{11} (2.13) to the bound channel (10). Therefore, according to the theoretical results of CHOW and THOMPSON /5.6/, the beams (0 ±1) can show intensity maxima at the resonance angles, as is actually found. The adjacent (-1 ±1) beams which are not directly coupled to the bound channel (v_{12} is vanishing small) indeed do show intensity minima. Expressed in a pictorial way, this result means that atoms which were scattered into the bound state ε_ν by a diffraction corresponding to the (01) undergo a second diffraction of type (-1 ±1) and reappear in the (0 ±1) beam /1.9/. Analogous observations were reported by FRANKL et al. /5.11/ for He scattering from LiF(100).

5.3 Theory of Atom Scattering from a Corrugated Hard Wall with an Attractive Well

Numerous theoretical attempts can be found in the literature aiming at solving the full problem of particle scattering from solid surfaces, whereby effects due to the attractive part of the potential as well as the interaction with lattice vibrations were considered /3.2;5.1,4,6,12-19/. The relation between the different methods proposed has been discussed in an article by WOLFE et al. /5.20/. Instead of surveying all these different approaches and their relative merits, we find it more useful for the present purpose to restrict ourselves to a detailed description of a rather recent formulation of the problem developed by CELLI et al. /5.21/, which has the advantages of being mathematically closely related to the formalism outlined in Chap.3, of being physically transparent, and of allowing a convenient numerical parameterization. Thus, it permits quantitative calculations for real systems, provided the hard-wall corrugation function $\zeta(\underline{R})$ and the bound-state energies ε_ν are known. Indeed, the method has already been applied with considerable success in describing resonant effects in atom diffraction from ionic crystals /5.22,23/, graphite /5.24-27/, and the adsorbate system Ni(110) + H(1×2) /5.28/.

CELLI et al. /5.21/ start from a simple model potential including a short-range repulsion V_r due to a hard wall described by $\zeta(\underline{R})$ and a long-range attraction V_a, as shown in Fig.5.7. There are two planes $z = z_0$ and $z = z_0 - \delta$ ($\delta > 0$), so that

$$V_r[z - \zeta(\underline{R})] \quad \begin{array}{l} = \infty \text{ for } z \leq z_0 - \delta \\ = 0 \text{ for } z > z_0 - \delta \end{array} \qquad (5.10)$$

and

$$V_a(z) = -D \text{ for } z < z_0 \quad . \qquad (5.11)$$

Thus, the attractive well has a flat bottom of depth D and a width > δ. The choice of such a potential has the consequence that the wave functions for V_r and V_a can first be found separately and that they can then be matched at $z = z_0$. In the re-

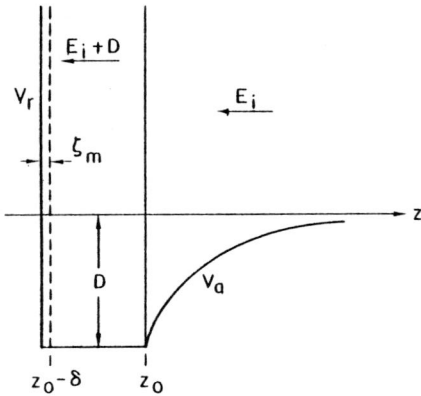

Fig.5.7. Model potential for hard corrugated wall with attractive wall used by CELLI et al. /5.21/

gion $z_0 - \delta < z < z_0$, the particle wave function has the form

$$\psi(\underline{r}) = \sum_{\underline{G}} B_{\underline{G}}^+ \exp i\left[(\underline{K}+\underline{G})\underline{R}+k_{\underline{G}z}'z\right] + B_{\underline{G}}^- \exp i\left[(\underline{K}+\underline{G})\underline{R}-k_{\underline{G}z}'z\right] \qquad (5.12)$$

with $k_{\underline{G}z}'$ given by (5.4). The far-field solution has the usual form [compare (3.15)]

$$\psi(\underline{r}\to\infty) = \exp i\left[(\underline{K}\underline{R}+k_{iz}z)\right] + \sum_{\underline{G}} A_{\underline{G}} \exp i\left[(\underline{K}+\underline{G})\underline{R}+k_{\underline{G}z}z\right] \qquad . \qquad (5.13)$$

The scattering amplitudes $A_{\underline{G}}$ are the quantities desired.

For the region of the potential $V_r - D$, the incoming waves are of the form $\exp i\left[(\underline{K}+\underline{G})\underline{R}-k_{\underline{G}z}'z\right]$ and their amplitudes $B_{\underline{G}}^-$ are related to the amplitudes $B_{\underline{G}}^+$ of the diffracted waves by

$$B_{\underline{G}}^+ = \sum_{\underline{G}'} S(\underline{G},\underline{G}')B_{\underline{G}'}^- \qquad . \qquad (5.14)$$

With the corrugation function $\zeta(\underline{R})$ known, the amplitudes $S(\underline{G},\underline{G}')$ can be calculated by using the methods outlined in Chap.3 taking into account the modifications of Sect.5.1. The scattering problem for V_a is easily solved, as V_a does not exhibit lateral periodic modulation, and therefore, parallel momentum is conserved. This means that with the wave $\exp i\left[(\underline{K}+\underline{G})\underline{R}+k_{\underline{G}z}'z\right]$ incident from the left on V_a, only $\exp i\left[(\underline{K}+\underline{G})\underline{R}-k_{\underline{G}z}'z\right]$ and $\exp i\left[(\underline{K}+\underline{G})\underline{R}+k_{\underline{G}z}z\right]$ can be corresponding outgoing waves. Therefore, for $\underline{G} \neq 0$,

$$A_{\underline{G}} = T_{\underline{G}} B_{\underline{G}}^+ \qquad (5.15)$$

and

$$B_{\underline{G}}^- = R_{\underline{G}} B_{\underline{G}}^+ \qquad , \qquad (5.16)$$

with R_G and T_G the reflection and transmission coefficients for incidence from the left, which are related through

$$k_{\underline{G}z}|T_{\underline{G}}|^2 + k_{\underline{G}z}'|R_{\underline{G}}|^2 = k_{\underline{G}z}' \quad . \tag{5.17}$$

For $\underline{G} = 0$, one must also include the initial incoming wave $\exp i(\underline{KR}-k_{0z}z)$ and therefore,

$$A_0 = R_0' + T_0 \, B_0' \tag{5.18}$$

$$B_0^- = T_0' + R_0 \, B_0^+ \tag{5.19}$$

with R_0' and T_0' being the coefficients for incidence from the right, where $T_0'^2 = k_{0z}/k_{0z}'$ in accordance with (5.17). Combining (5.15,16,18,19) one obtains an infinite set of linear equations of the form

$$B_{\underline{G}}^+ = \sum_{\underline{G}} S(\underline{G},\underline{G}')R_{\underline{G}}' \, B_{\underline{G}'}^+ + S(\underline{G},0)T_0' \tag{5.20}$$

that, according to CELLI et al. /5.21/, can be restricted in most cases to the set of vectors $\{\underline{N}\}$ lying within the two Ewald spheres with radii k_i and k_i' (compare Fig.5.1). Thus, (5.20) becomes a matrix equation of finite (and usually small) size involving the unknowns $B_{\underline{N}}^+$. The $R_{\underline{N}}$ can be specified /5.21,26,27/ through a phase shift $\delta_{\underline{N}}$ that characterizes reflection of the particles from the attractive potential. $\delta_{\underline{N}}$ can be obtained easily from an interpolation of the experimental bound-state energies ε_ν if they obey approximately the relation

$$\varepsilon_\nu = -D\left[1-\Lambda(\nu+\tfrac{1}{2})\right]^\alpha \quad , \tag{5.21}$$

where the parameters D, α, and Λ fix the depth, range, and steepness (or asymmetry) of the potential, respectively /5.26/. The reflection amplitudes desired can be obtained by writing (5.21) as a function of ν,

$$\nu(\varepsilon) = \frac{1}{\Lambda}\left[1-\left(-\frac{\varepsilon_\nu}{D}\right)^{1/\alpha}\right] - \frac{1}{2} \tag{5.22}$$

through

$$\delta_{\underline{N}} = 2\pi\nu(\varepsilon_\nu) \tag{5.23}$$

and

$$R_{\underline{N}} = \exp\left[i\delta_{\underline{N}}\right] \quad . \tag{5.24}$$

Thus, it suffices to characterize the attractive part of the potential by the phase shifts δ_N, and therefore, the assumption of the potential described by (5.10,11) constitutes only a mathematical trick and does not restrict the applicability of the method to the peculiar shape of the potential shown in Fig.5.7.

The set of equations (5.20) can now be solved for the $B_{\underline{G}}^{+}$, and the total scattering intensities for the beams $\underline{F}$ within the Ewald sphere of radius k_i can be calculated according to

$$P_{\underline{F}} = \frac{k_{\underline{F}z}}{k_{0z}} |A_{\underline{F}}|^2 = \frac{k'_{\underline{F}z}}{k'_{0z}} |\sum_{\underline{N}} S(\underline{F},\underline{N})B_{\underline{N}}^{+} R_{\underline{N}}/T'_0 + S(\underline{F},0)|^2 \quad . \tag{5.25}$$

A detailed analysis of the conditions leading to minima, maxima, or Fano-type anti-resonance structures in the angular or energy dependence of the intensities can be found in the original paper /5.21/. As the theory does not take into account inelastic effects, the sum of the calculated intensities $P_{\underline{F}}$ obeys the unitarity condition (3.17).

The theoretical approach outlined above has been applied by GARCIA et al. /5.22/ to the case of He/LiF(100). Figure 5.8 shows, in the upper part, the result of the elastic calculation and, in the lower part, experimental data of FRANKL et al. /5.11/. The calculation was based on the best-fit corrugation function for LiF obtained by GARCIA /3.19/ (Chap.7) and the bound-state energies obtained by MEYERS and FRANKL /5.5/. A comparison shows that all the experimentally observed features are reproduced quite well, although all the calculated structures are sharper than the experimental ones. The theoretical intensity had to be multiplied by an overall factor of 0.43 to obtain agreement with the experimental intensity. This is due to inelastic losses in the experiment and corresponds to a Debye-Waller correction (Chap.4).

5.4 Inelastic Effects in Resonant Scattering

In their work on the He/LiF /5.22/, GARCIA et al. conjectured that the differences in the width and sometimes also the shapes of the resonant structures between the experimental and theoretical results might be due to inelastic effects. Their conjecture was corroborated by theoretical results concerned with He scattering from graphite(0001) /5.24/, as inelastic effects should play a larger role in graphite than in LiF because of its smaller Debye temperature. Figure 5.9a shows the experi-

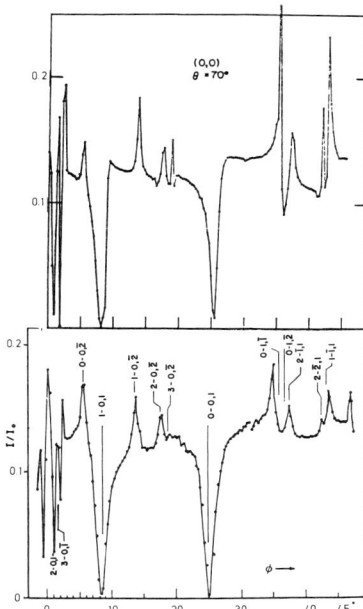

Fig.5.8. Elastic theory (upper part) and experiment (lower part) for the azimuthal dependence of the specular intensity θ_i = 70° for the He/LiF(100) at k_i = 5.76 $\overset{o}{A}^{-1}$. The theoretical intensities are multiplied by a factor 0.43. Energy levels and resonant channels are identified in the experimental plot /5.22/

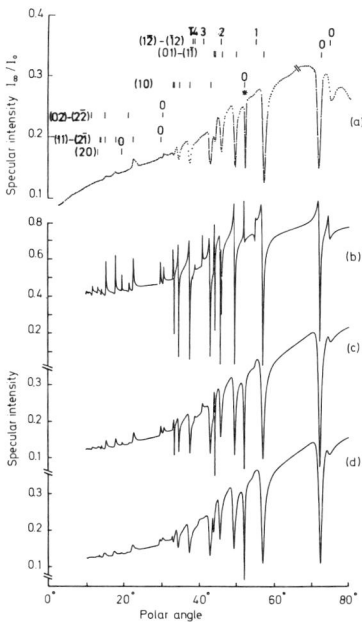

Fig.5.9a-d. Specular scattering of 22 meV helium atoms from the (0001) surface of graphite. (a) Experimental results of BOATO et al. /2.9/, (b) elastic theory, (c) with Debye-Waller attenuation according to (5.27), (d) Debye-Waller attenuation plus convolution with the energy dispersion of the experimental beam. Energy levels and resonant channels are labelled in (a). The asterisk indicates a channel strongly coupled only to the specular beam that changes from a maximum in elastic theory to a minimum when Debye-Waller attenuation is included /5.26/

mental result of BOATO et al. /2.37/ for the angular dependence of the specular beam. In Fig.5.9b the result obtained with the elastic theory /5.24/ is shown. The calculations were based on the corrugation function obtained by BOATO et al. /2.36/ (compare Chap.8) and the bound-state energies cited in the same publication. Comparison of Figs.5.9a,b reveals large deviations of the purely elastic theory from experiment concerning both the intensity and the details of the resonance structures. Note in particular that the observed widths of the resonance features are a factor of three larger than those of the elastic calculation. The structure marked by an asterisk in Fig.5.9a is observed as a minimum, whereas the elastic calculations give a maximum. An analysis of the kinematic conditions (5.5) shows that in this case the bound state is coupled strongly only to the specular and to no other diffraction channel. This corresponds to a situation where the Weare-Wolfe rule 2 is applicable (Sect.5.2).

As the intensities obtained from elastic theory did not adequately reproduce the resonant structures observed for He/graphite, HUTCHISON et al. /5.25/ proposed incorporating the Debye-Waller factor into the theory outlined in Sect.5.3 by making the following replacement for *all* hard-wall amplitudes,

$$S(\underline{G},\underline{G}') \rightarrow \exp\left[-W(\underline{G},\underline{G}')\right] S(\underline{G},\underline{G}') \quad , \tag{5.26}$$

where the exponent in full analogy to (4.2),

$$W(\underline{G},\underline{G}') = \frac{1}{2} <\rho_z>^2 \left[k_{\underline{G}z}' + k_{\underline{G}'z}'\right]^2 \quad , \tag{5.27}$$

contains the perpendicular momentum transfer with the well correction of Sect.5.1 included and the mean-square displacement of the surface atoms perpendicular to the surface $<\rho_z>^2$. As outlined in Chap.4, for the scattering from a hard surface where the collision process is of short duration compared to a typical phonon frequency, the Debye-Waller correction of the form (5.27) holds for both completely correlated and uncorrelated motions of the surface atoms /4.4/.

The result of this rather straightforward introduction of the Debye-Waller factor for the case He/graphite is shown in Fig.5.8c /5.26,27/. Now, the overall intensities as well as the shapes and widths of the bound-state resonances are in remarkable agreement with experiment. The agreement becomes even more convincing when the finite-energy spread of the experimental beam is also taken into account (Fig. 5.8d). Note that the resonant transition marked by an asterisk in Fig.5.8a, which produces a maximum in purely elastic theory, is changed in the inelastic theory to a minimum in agreement with observation (for a detailed discussion of when and why this can happen, see /5.21,25-27/).

As mentioned at the end of Sect.5.3, in the purely elastic theory, unitarity is conserved. This is no longer the case when the Debye-Waller correction is taken into account according to (5.26). An experimental and theoretical investigation of this situation has been performed recently by GREINER et al. /5.23/. Figure 5.10 shows the result of a D-scattering experiment from LiF(001), which corresponds to

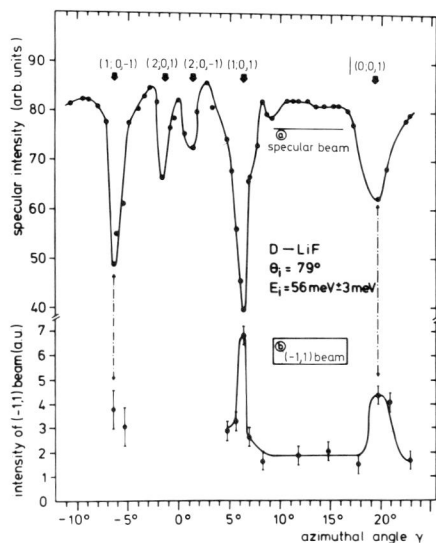

Fig.5.10. Selective adsorption resonances of D_1/LiF(100) appearing in the specular and the $(\bar{1}1)$ beams when azimuthal orientation γ is varied. Only part of the intensity disappearing in the (00) reappears in the $(\bar{1}1)$ /5.23/

strong coupling of the $(\bar{1}1)$ beam to the resonant channel (01) through the strongest periodic term in the potential v_{10}. Therefore, at $\gamma = 6°$, a minimum appears in the specular, whereas a maximum is observed for the $(\bar{1}1)$ beam. However, only part of the intensity disappearing in the (00) reappears in the $(\bar{1}1)$. According to the discussion at the end of Sect.5.2 (Fig.5.6), this can be easily understood as the particles in the bound state moving for an appreciable time parallel to the surface and therefore being able to suffer several inelastic events before being released from the surface. Calculations on the basis of the inelastic theory showed that this intensity loss is reproduced at least semiquantitatively /5.23/.

In a further recent study, HUTCHISON and CELLI /5.27/ have shown that although band splitting is described by the theory outlined in Sect.5.3, a periodic modulation of the well depth D has to be included to reproduce the observed splitting for the He/graphite system /2.8,9,37/ properly.

Finally, we mention the work of SOLER et al. /5.28/, who have shown for the case of He scattering from the adsorbate system Ni(110) + H(1×2) that the shape of the resonance can be used to decide whether $+\zeta(\underline{R})$ or $-\zeta(\underline{R})$ describes the real surface profile. In this case, the intensity analysis on the basis of the eikonal approximation has left this question open (compare Sect.3.4.2). For small corrugations, the scattering amplitudes for $+\zeta(\underline{R})$ and $-\zeta(\underline{R})$ are nearly complex conjugate to each other, so that they lead to almost the same intensities in cases of nonresonant scattering (3.16). On the other hand, complex conjugation of the amplitudes has a strong influence on resonant structures as is easily seen from (5.25). Therefore, a decision which sign is the correct one is possible by analyzing resonant line shapes.

6. Experimental Aspects of Gas-Surface Scattering

6.1 Requirements on an Apparatus to Perform Gas-Surface Scattering Experiments

The requirements placed on an apparatus to perform diffraction experiments on single-crystal surfaces are numerous. Since it is described to carry out studies on reactive surfaces such as metal and semiconductors, the scattering chamber should be capable of a base pressure of better than 1×10^{-10} Torr. The surface structure and chemical composition should be known, which necessitates including several of the standard surface-analysis techniques such as LEED, AES, UPS, XPS, EELS, SIMS, and ion scattering. Most materials can be cleaned only by a combination of heating cycles, ion bombardment, and chemical reaction, so that a sputtering gun and gas-inlet facilities should be present in the apparatus. Since the angle between the beam and the surface normal should be variable, the sample manipulator should allow angles of rotation about two perpendicular axes, one of which is parallel to the surface normal. This assumes that the detector can be rotated out of the plane defined by the beam direction and the surface normal. If the detector can be rotated only in this plane, the manipulator should allow rotations about three mutually perpendicular axes, one of which is parallel to the surface normal.

Apart from the need that the substrate be cleaned, characterized, and its orientation relative to the beam changed, additional requirements are placed on the beam source and the detector for the scattered particles. The beam should have an intensity at the surface of approximately 10^{14}-10^{15} particles cm^{-2} s^{-1} and should be as monoenergetic as possible for diffraction studies. The angular divergence of the beam and its diameter at the surface should be small to increase the angular resolution attainable. The detector should also have a small aperture and a selectively high sensitivity for the beam gas. As UHV techniques /6.1,2/ and surface-analysis methods /6.3,4/ have been discussed elsewhere, we shall restrict our review of experimental aspects to beam sources, beam detectors, and sample manipulators suitable for surface studies. In the final section we shall discuss the transfer widths of molecular beam systems and the effect of crystalline imperfections on the measured intensities. In discussing beam sources, we have drawn on earlier reviews /6.5-8/ for much of the following material.

6.2 Beam Sources

6.2.1 Effusive Beam Sources

The simplest sources which can be constructed are of the effusive type. From elementary kinetic theory /6.9/, the number of particles which emerge from a thin orifice of area A_s at an angle θ to the normal of the orifice plane into a solid angle $d\omega$ is given by

$$dI(\theta,r) = \frac{d\omega}{4\pi} n\bar{v}A_s \cos\theta \quad , \tag{6.1}$$

where n is the density of the gas, and $\bar{v}$ is the mean molecular velocity. In terms of source pressure P_0 in Torr, temperature T, and molecular weight m, (6.1) becomes

$$I(\theta,v) = 1.1\times10^{20}\, P_0 A_s \cos\theta (mT)^{-1/2}\ \text{particles sr}^{-1}\,\text{s}^{-1} \quad . \tag{6.2}$$

Both (6.1,2) are valid only for source pressures such that the mean free path behind the orifice λ is greater than or equal to the smallest dimension of the orifice d. Because of the low density, $I(\theta,v)$ will be to first order independent of the gas used and is 5×10^{16} particles $\text{sr}^{-1}\,\text{s}^{-1}$ at 300 K. At a distance of 1 meter from the source, this corresponds to a beam intensity of 5×10^{12} particles $\text{cm}^{-2}\,\text{s}^{-1}$.

Two disadvantages of effusive sources for diffraction studies are the Maxwellian velocity distribution of the beam and the low beam intensity integrated over all velocities given by (6.2). Since a monoenergetic source is necessary for diffractive investigations (otherwise, in general only first-order diffraction can be resolved), it is necessary to combine the effusive source with a velocity selector, which adds to the system complexity. With slotted disc selectors /6.10,11/, a sufficiently monoenergetic beam can be generated, but only with an appreciable loss in intensity since the velocity distribution of the effusive source is broad. The poor directivity of the single orifice effusive source can be improved by using long capillary arrays /6.12,13/. However this increase in directivity is most pronounced for low source pressures which yield low intensity beams. As the source pressure is increased, the directivity becomes comparable to that of a simple orifice source.

6.2.2 Nozzle-Beam Sources

Historically, JOHNSON /6.14/ showed using Hg beams that the intensity increased even as the Knudsen number K_n ($K_n = \lambda/d$) was decreased far below unity. Their initial experiments showed that the detector signal increased linearly with source pressure up to values for K_n of 10^{-3}. This corresponds to continuum flow in the orifice region which is followed by a transition to molecular flow as the beam density decreases far from the orifice. KANTROWITZ and GREY /6.15/ first proposed the use of continuum sources to achieve high beam intensities, and the first experimental realizations were by KISTIAKOWSKY and SLICHTER /6.16/ and BECKER and BIER /6.17/. In their designs, a converging-diverging nozzle was combined with a sharp-edged conical skimmer whose function is to avoid a detached shock wave in front of the skimmer. The sharp edge and shape of the skimmer also minimize interference effects which can occur from molecules scattered from its inside and outside surfaces. Later work has shown that the nozzle form is not important for most purposes, and simple free jets have been used by most workers in recent studies. The skimmer quality is an important factor in designing high-intensity highly monoenergetic beams, and design criteria /6.18/ and interference effects /6.19/ have been discussed elsewhere. A free jet source can be simply fabricated by welding or press-fitting commercially available electron-microscope apertures into a suitable holder, and skimmers have also become commercially available /6.20/ in recent years.

The gas dynamics of free jet sources has been studied by a number of investigators /6.21-26/. In the original study of KANTROWITZ and GREY /6.15/, the velocity distribution at the skimmer entrance is given by

$$I(u,v,w)dudvdw = \left(\frac{m}{2\pi kT_s}\right)^{3/2} \exp\left\{-\frac{m}{2kT_s}\left[(v-v_s)^2 + u^2 + w^2\right]\right\} dudvdw \quad , \tag{6.3}$$

where v is the particle velocity parallel to the beam axis, u and w are the perpendicular components, and T_s is the gas temperature at the skimmer entrance. This differs from the velocity distribution from an effusive source in that v_s is nonzero and in that T_s is the gas temperature at the skimmer rather than the orifice temperature as for the effusive source. Since T_s is much lower than the skimmer temperature, the velocity distribution is narrow and is centred near the flow velocity v_s. This can be seen in Fig.6.1, in which M_s is the Mach number of the beam. M_s is defined as the ratio of u_s to c_s, where c_s is the local sound velocity given by $(\gamma T_s/m)^{1/2}$, and γ is the specific-heat ratio C_p/C_v. Assuming an isentropic expansion, the maximum in the velocity distribution for a nozzle beam is greater than that for an effusive source by the factor $[2\gamma/3(\gamma-1)]^{1/2}$, which is 1.29 for a monoatomic gas such as He.

In recent years, it has become more common to characterize the velocity distribution of a nozzle beam by the speed ratio S given by

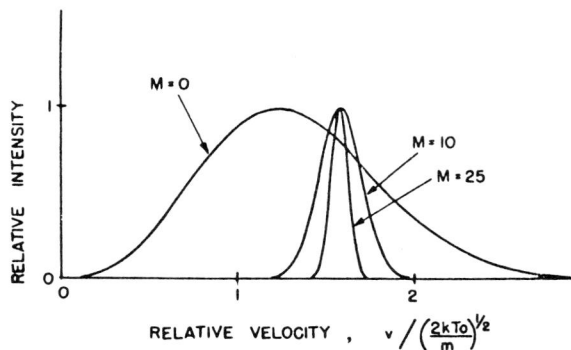

Fig.6.1. Theoretical velocity distributions of effusive and nozzle-beam sources /6.27/

$$S = \left(\frac{1}{2} mv_s^2/kT \right)^{1/2} \qquad (6.4)$$

where T is the temperature describing the mean kinetic energy of the molecular motion in the gas which moves with the flow velocity v_s. For large speed ratios, $\Delta v/v_s$ and T/T_0 (where Δv is the full width at half maximum (FWHM) of the velocity distribution about the average value v_s, and T_0 is the temperature of the stationary gas prior to the expansion) are given by /6.26/

$$\Delta v/v_s \sim 1.65/S \qquad (6.5)$$

and

$$T/T_0 \sim 2.5/S^2 \quad . \qquad (6.6)$$

Speed ratios as high as 350 have been reported for He beams /6.28/, whereas much lower limiting values are obtained for gases such as Ne due to beam condensation at high values of P_0.

The speed ratios which can be attained depend on the stagnation pressure P_0 and the nozzle diameter d as can be seen in Fig.6.2 which shows calculated values for He beams. The assumption of an isentropic expansion even at large distances from the nozzle leads to speed ratios independent of P_0d. However, this assumption does not correctly take into account that no further cooling takes place after the transition from continuum to molecular flow has taken place. As can be seen from the results in Fig.6.2, for increasing values of P_0d the distance at which this transition takes place is further from the nozzle, allowing the expansion to reduce the temperature T more, leading to an increase in S. Since the total flow through the nozzle is proportional to P_0d^2, large values of S can best be achieved for a given throughput using small nozzle diameters. However, nozzle clogging by impurities sets a practical lower limit of about 3×10^{-2} mm for d.

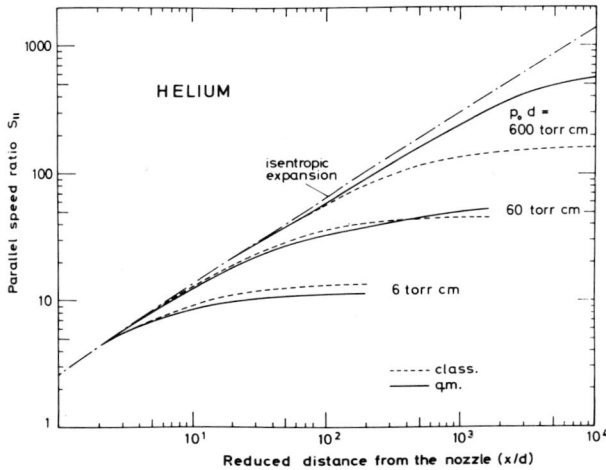

Fig.6.2. Calculated parallel speed ratios based on classical (- - -) and quantum mechanical (————) calculations. Also shown is the calculated curve for an isentropic expansion (——·——·——·) /6.26/

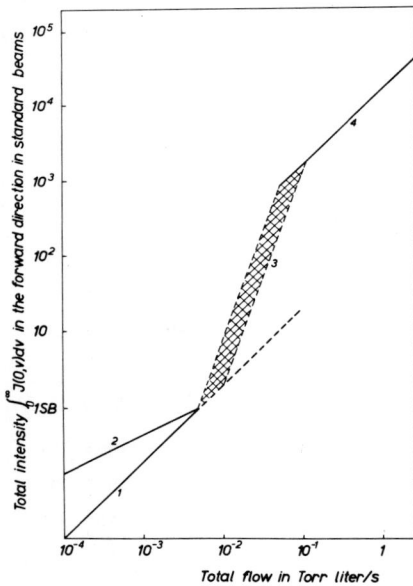

Fig.6.3. Theoretical performance of several beam sources. The velocity integrated beam intensity is shown as a function of the total nozzle throughput. (1) refers to a molecular effusion source, (2) to a multichannel array with the same total area as (1), (3) to the transition region from molecular to hydrodynamic flow, and (4) to a hydrodynamic flow source. The nozzle area is the same as (1), $M = 10$, $\gamma = 1.4$ /6.8/

The advantages of nozzle sources when compared with effusive sources for diffractive gas-surface investigations lie in the high intensity coupled with the high degree of monoenergicity which can be achieved without velocity selection. A comparison of the velocity integrated intensity obtainable with effusive, multichannel effusive, and nozzle sources is shown in Fig.6.3. One standard beam is defined as the maximum obtainable effusive intensity which is 5×10^{16} particles $sr^{-1} s^{-1}$. It is seen that the total available intensity is much higher for a nozzle source, but at the expense of a much increased throughput. The advantage of a nozzle source for diffraction experiments becomes even clearer when it is considered that the effusive source intensity in Fig.6.3 must be reduced considerably, since velocity selection is necessary before impingement on the surface.

6.3 Beam Energy Variation for Effusive and Nozzle-Beam Sources

For effusive sources the energy can be varied by heating and cooling the source. The most probable velocity is given by $(3kT_0/m)^{1/2}$, and the velocity distribution, by (6.3) with $v_s = 0$ and $T_s = T_0$. The lower limit is set by the vapor pressure of the gas being used, whereas the upper limit is set by the mechanical properties of the oven. If refractory metals such as tungsten can be used, temperatures in excess of 2000 K can be reached. The energy of nozzle sources can be varied in the same way, although the most probable velocity is shifted to higher values by the factor $[2\gamma/3(\gamma-1)]^{1/2}$, and the nozzle cannot be cooled to very low temperatures, since condensation can occur in the expansion. The speed ratio will decrease as T_0 increases, but a simple relationship valid for all gases cannot be derived.

Additional methods are also available to extend the energy range for nozzle sources to higher and lower values. One of these techniques is seeding /2.29/, in which a seed gas is added in small quantities (less than 10%) to a carrier gas. The seed-gas particles are swept through the nozzle as if they were carrier-gas particles and, depending on the relative molecular weights of the seed and carrier gases, will be accelerated or decelerated. Very high energies can be achieved with arc sources /6.30/, but this energy range is not useful for diffraction experiments.

For He beams, the energy range available by heating and cooling the nozzle is sufficient for diffraction experiments on surfaces. The de Broglie wavelength corresponding to the most probable velocity of a nozzle beam of mass m and source pressure T_0 is given by

$$\lambda = 19.58 \ (mT_0)^{-1/2} \quad . \tag{6.7}$$

Nozzle sources with T_0 as low as 3 K have been reported /6.31/, and temperatures of 2000 K should be accessible. This corresponds to a range from 0.2 to 5.6 Å which is more than sufficient for surface-diffraction studies.

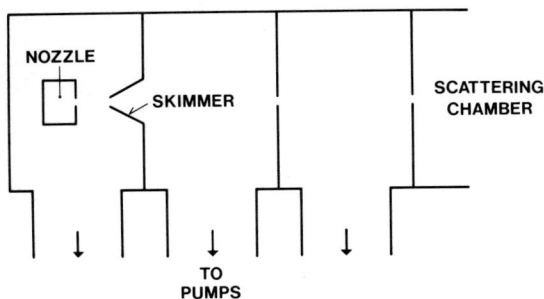

Fig.6.4. Schematic drawing of a nozzle-beam system consisting of a nozzle-skimmer chamber, several differential pumping stages, and a scattering chamber

6.4 The Design of Nozzle-Beam Systems

A schematic drawing of a nozzle-beam system is shown in Fig.6.4. It consists of a nozzle-skimmer chamber and a number of differentially pumped chambers intermediate between the nozzle-skimmer and UHV scattering chambers. The intermediate chambers are necessary to reduce the pressure from the high value in the nozzle-skimmer chamber to a value compatible with UHV investigations and to insure that the flux into the scattering chamber is almost entirely due to the direct beam. The number of differential chambers needed depends on the pressure in the nozzle-skimmer region, the beam diameter, and the pumping speeds in the differential stages. A simple design criterion is that the effusion rate from the last differential stage calculated using (6.1) should be much smaller than the beam flux.

Conventional nozzle-beam systems for surface studies have been pumped in all stages with baffled oil-diffusion pumps, which limit the nozzle-skimmer region pressure to at most 10^{-3} Torr. At these pressures, collisions between the beam and background particles can substantially reduce the speed ratio and beam intensity. At a throughput of 1 Torr l s^{-1}, which for helium beams corresponds approximately to P_0d values of 10 and a speed ratio of 10, pumps with speeds of 10^3 l s^{-1} above the baffle are required. Such pumps require flange openings of 200-300 mm diameter, resulting in rather bulky and expensive systems. Since more than 99% of the nozzle throughput in a typical system is evacuated in the nozzle-skimmer chamber, the subsequent pumps can be substantially smaller if several differentially pumped stages are included. However, because S rises with P_0d to a power of roughly 0.4 /6.27/, the generation of highly monoenergetic beams requires very large pumping speeds in the nozzle-skimmer chamber.

An alternative design for a nozzle-beam system has been proposed and realized by CAMPARGUE /6.28,32,33/. Rather than eliminating background-beam collisions by decreasing the nozzle throughput, in this design the jet density is increased to the point that a barrel shock is formed about the free jet which shields the beam from collisions with the background gas. Due to the high-pressure tolerance in the nozzle-skimmer region of 0.25 Torr, P_0d values of up to 500 Torr cm have been used with resulting speed ratios of 350 for helium /6.28/. In this pressure range, high throughputs can be pumped through relatively small flange diameters with Roots or oil-ejector pumps, which are much less expensive than comparable throughput oil-diffusion pumps operating in the 10^{-3}-10^{-4} Torr range.

Another recent advance in nozzle-beam systems is the development of sources which can be pulsed on for periods as short as 10 µs /6.34/. If the time between successive pulses is much longer than the pulse width, a high instantaneous beam intensity can be achieved with a low average throughput. Such a design requires much smaller pumps than are needed for a continuous source. However, such sources are limited to applications such as inelastic scattering studies in which a low duty cycle does not necessarily lead to a corresponding increase in the measuring time.

In closing this section, a few examples of nozzle constructions suitable for gas surface-scattering studies will be presented. The construction used in the authors' laboratory for He, H_2 and Ne beams is shown in Fig.6.5. The source is a 1 mm O.D. platinum tube, in whose side a hole of $\sim 8\times10^{-2}$ mm has been spark eroded. The tube is clamped between massive copper supports, which can be cooled with water or liquid nitrogen. The temperature of the tube can be regulated to better than 1 K between 80 K and 1500 K by direct current heating, whereby the thermoelement output is used to drive the temperature regulator. The speed ratio can be substantially improved by using smaller orifice sizes. Other designs have been published which allow either heating /6.35/ or cooling /6.36/ to extremely high or very low temperatures.

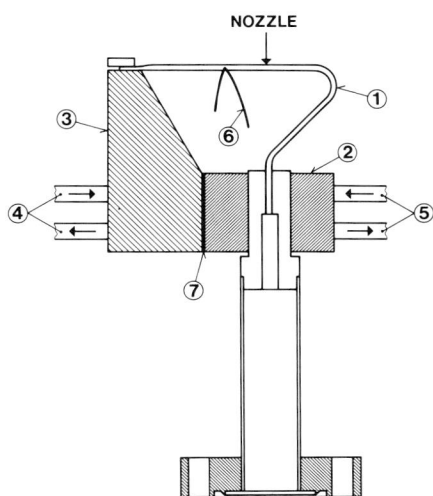

Fig.6.5. Schematic drawing of the variable-temperature nozzle source used in the authors' laboratory. (1) platinum tube in which the nozzle is spark eroded, (2) and (3) copper supports, (4) and (5) copper tubing for liquid-nitrogen cooling. The heating current also passes through the tubing. (6) thermoelement, (7) teflon insulator

The only particles other than He, H_2, and Ne for which diffraction from surfaces has been observed are atomic hydrogen and deuterium. Although thermal sources can be constructed easily /6.37/, most operate in the effusive mode, which necessitates their use together with a velocity selector for diffraction studies. However, super-sonic thermal H sources have also been reported /6.38/ with a lower degree of dis-sociation than would be obtained at the same temperature for lower pressures, but with a more desirable velocity distribution for diffraction experiments. Microwave discharge sources have been successfully used to generate supersonic beams of atomic species, and one example of such a source /6.39/ is shown in Fig.6.6. Source pres-sures of up to 200 Torr have been used in this design.

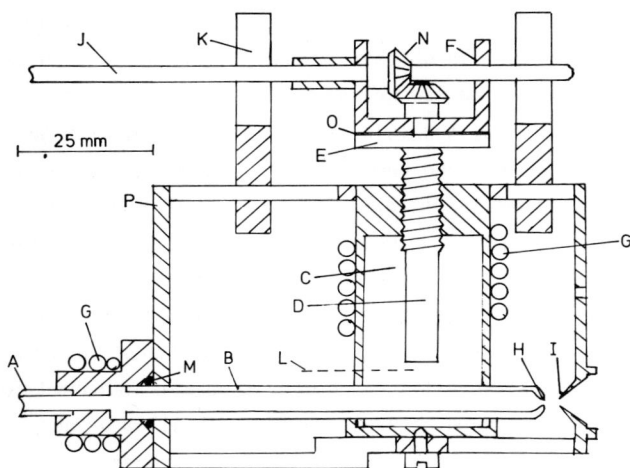

Fig.6.6. Schematic drawing of a microwave discharge source. (A) gas-inlet tube, (B) quartz discharge tube, (C) microwave cavity, (D) tuning electrode, (E) sup-port disc, (F) gear housing, (G) water-cooling tube, (H) nozzle, (I) skimmer, (J) drive shaft, (K) shaft guides, (L) axis of coupling stab, (M) O-ring seal, (N) level gears, (O) teflon washer, (P) framework /6.39/

6.5 Molecular-Beam Detectors

Molecular-beam detectors should have a high selective sensitivity for the beam gas to increase the signal-to-noise ratio and to reduce the measurement time necessary for a diffraction trace to a minimum. This is particularly important for highly re-active surfaces for which the measurement time is limited to about 30 minutes to avoid surface contamination even at reactive gas partial pressures of 1×10^{-10} Torr.

Most investigators have used ionization detectors combined with mass spectrometers to achieve the selective sensitivity required. The detectors can be either
of the flow-through type /6.40-42/, in which the density is measured, or of the
stagnation type /6.43/, in which the flux is measured. If the apparatus is to be
capable of time-of-flight measurements in addition to diffraction experiments, only the flow-through detector can be used. For diffraction experiments, a choice between these two types should be based on the sensitivity, time constant, and ease
of construction of the detector. For flow-through ionization detectors which are
not differentially pumped, a sensitivity of 4×10^{-4} I_0, where I_0 is the direct beam
intensity, has been obtained in the authors' laboratory with a commercially available mass spectrometer. (Phase-sensitive detection at 75 Hz, 1:1 signal-to-noise
ratio, 1 s time constant.) Stagnation detectors in similarly designed systems have
been able to detect 10^{-3} I_0 /6.44/. The time constant τ of the stagnation detector
is given by V/S_p, where V is the enclosed volume about the detector, and S_p is the
conductance of the aperture. Although τ in many applications is as high as 30 seconds
/6.43/ for small detector volumes, τ can be made sufficiently low that stagnation
detectors can be combined with chopped beams and phase-sensitive detection to yield
better signal-to-noise ratios than conventional flow-through detectors /6.45/. The
sensitivity limits cited above are also dependent on the background pressure at the
detector and can be substantially increased either by differentially pumping the
detector or by incorporating extremely high-speed pumps in the scattering chamber.
Sensitivities of 1×10^{-5} I_0 /6.46/ and better than 10^{-4} I_0 /4.17/ have been obtained
with a single differential pumping stage around the detector and with a cryogenically pumped system and stagnation detector, respectively. For well-ordered surfaces
with not too many intense diffraction beams, detectors which are not differentially
pumped are sufficient for diffractive studies. Figure 6.7 shows a He-diffraction
trace from the (1×2) reconstructed Au(110) surface, which was obtained with a time
constant of 0.3 s in less than 180 s. Such a rapid rate of data accumulation is
often necessary as in the above case in which 50 such spectra at different angles
of incidence were measured in a 2-hour period to analyze resonant scattering from
the Au-He potential well. Surface contamination would have falsified the results if
a substantially longer measurement time had been used.

Another type of detector, which has been used particularly in combination with
cryogenically pumped systems, is the bolometer. This detector is sensitive to the
energy of the impinging particles rather than to the flux or density. Investigations using bolometers in the diffraction experiments have achieved a sensitivity
of better than 10^{-4} I_0 /2.11/. Recently, a superconducting bolometer, whose sensitive element is a tin or indium film, has been described /6.47/ which has a response
time of 1 μs. The minimum detectable signal (signal-to-noise 1:1, 1 s time constant)

He Diffraction from
Au (110) (1×2)

T_s = 100 K
λ = 1.09 Å
θ_i = 25°
[001] azimuth

Scattered intensity

Scattering angle θ

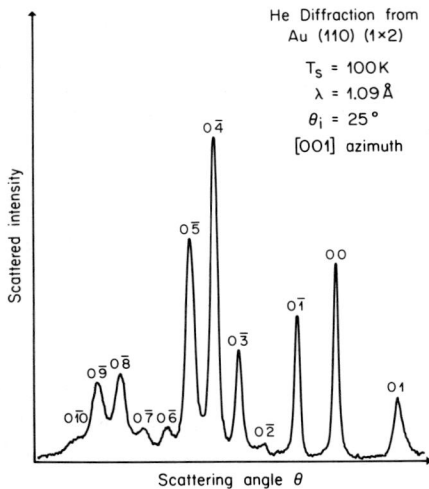

Fig.6.7. Diffraction scan from the Au(110) (1×2) surface illustrating the signal-to-noise ratio for a rapid scan with a time constant of 0.3 s

is 1.2×10^8 Ar atoms s^{-1} in a detector area of 4×10^{-2} cm^2. The fast response time also allows such bolometers to be used for time-of-flight measurements.

Atomic hydrogen can best be detected by monitoring the change in the surface conductivity of a ZnO crystal upon H adsorption /6.48/. The detector is highly selective and responds only to atomic hydrogen and oxygen. It is linear with hydrogen coverage over a wide range and can be periodically regenerated by desorbing the adsorbed atoms by a mild heating. The minimum detectable signal is about 10^{11} atoms cm^{-2} s^{-1}.

Other detector modifications which have been described include a cold-cathode ion source, which enables a quadrupole mass spectrometer to be used in a cryogenic environment /6.49/ and a field ionization detector for helium atoms /6.50/.

6.6 Detector Rotation

In diffraction experiments, the intensity of as many possible diffraction features should be measured for fixed angles of incidence of the molecular beam. This requires rotation of the detector about the scattering centre. One approach has been to rotate the detector within an all-metal sealed vacuum system using rotary-motion feedthroughs to transmit the motion through the chamber walls.

The design used in the authors' laboratory is of this type and is shown schematically in Fig.6.8. It allows for two independent rotations, one of which is in the scattering plane and the second perpendicular to this plane. The bearing for the in-plane rotation consists of an upper plate and a lower ring with a V-shaped circular groove. Three equally-spaced carbide balls of 15 mm diameter separate the

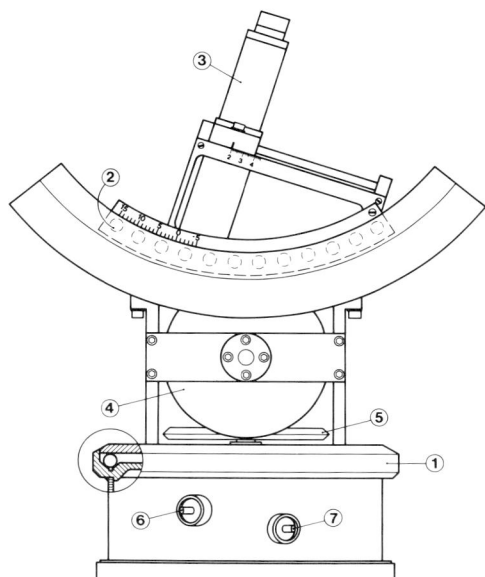

Fig.6.8. Schematic drawing of the detector rotation mechanism used in the authors' laboratory. (1) bearing for in-plane rotation, (2) bearing for out-of-plane rotation, (3) quadrupole mass-spectrometer detector, (4) and (5) coupling gears, (6) and (7) drive shafts for in-plane and out-of-plane rotations

rings, and allow low friction rotation without a bearing cage /6.51/. The bearing for the out-of-plane rotation is similar in design but contains a perforated ring segment to keep the balls equally spaced. The angular motion requires flexible detector cables which in our design limit the motion to 200° in plane as well as 15° above and 40° below the scattering plane. Although this design has the important advantage that out-of-plane scans can be recorded, it cannot easily be combined with a differentially-pumped detector.

A second approach is to fix the detector on a large flange and to rotate the entire flange as has been done by AUERBACH et al. /6.52/. Their design is shown schematically in Fig.6.9. The seals are made with spring-loaded MoS_2 filled teflon O-rings. Due to the differential pumping between adjacent O-rings, the static leak rate is less than 10^{-13} Torr l s^{-1}. The pressure rise while rotating the flange is less than 1×10^{-10} Torr. This design allows for convenient differential pumping of the detector, but must be combined with a three-axis manipulator in order to measure out-of-plane scattering.

6.7 Sample Manipulators

Examples of two- and three-axis manipulators suitable for use in UHV will be presented in this section. In order to carry out He-diffraction studies, it would be desir-

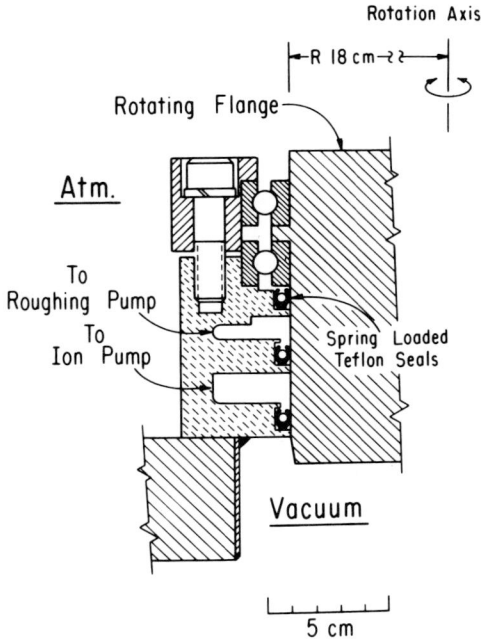

Fig.6.9. Schematic drawing of a rotating flange for use in UHV showing the spring-loaded teflon seals and the differential pumping stages /6.52/

able to modify these or other similar designs to be able to reach as low temperatures as possible. In this way, the uncertainty in correction for inelastic effects (Chap.4) can be minimized.

The two-axis manipulator shown in Fig.6.10 /6.53/ features an electrical readout for the azimuthal angle based on a wire-wound potentiometer in vacuum. The mechanism can be added onto commercial manipulators which have a flip actuator. The polar angle can be varied by ±200°, and the azimuthal angle variation is ±100° with an electrical resolution of 0.5°. The construction is completely nonmagnetic, and the eight adjustment screws allow a 5 mm linear displacement of the sample as well as a 5° tilt to align the crystal surface parallel to the bearing plane.

The three-axis manipulator shown in Fig.6.11 /6.54/ allows independent angular variations of 240°, 360°, and 40° about the A, B, and C axes, respectively, with an angular resolution of 0.02°. The rotation about the A axis is carried out with the central shaft of the manipulator. A rotation of one of the outer two of the three coaxial drive shafts causes a keyed lead nut to translate along the threaded lower end. The movement is converted back to a rotational movement by means of a rack and pinion. The sample is mounted in a gimbal assembly which allows it to rotate about its normal axis regardless of the tilt setting. The high angular resolution is achieved by spring loading the keyed nuts to eliminate backlash and by using large gearing-down ratios of 360:1 for the B and C rotation and 100:1 for the A rotation.

SIDE VIEW   FRONT VIEW

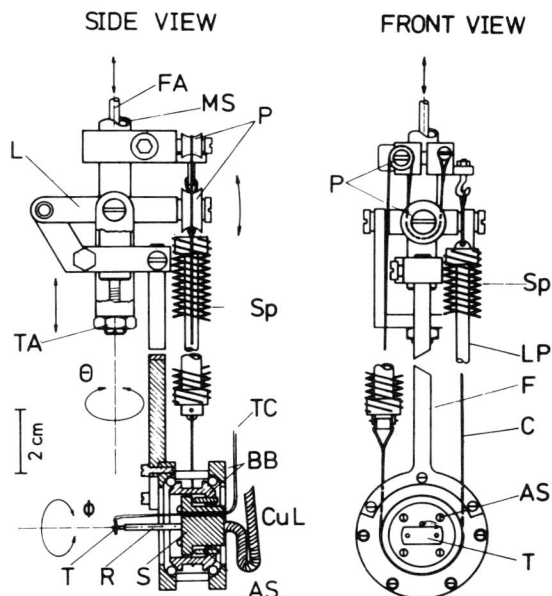

Fig.6.10. Drawing of the manipulator attachment for azimuthal rotation with electrical readout. (T) target, (R) W rod, (S) alumina support, (AS) adjustment screws,
(CuL) copper current leads, (BB) ball bearings with sapphire balls, (TC) thermo-
couple, (F) holder, (C) transmission cord, (MS) manipulator shaft, (FA) flip actuator, (L) lever, (TA) travel adjustment, (P) pulleys of the reversed block and
tackle, (Sp) return spring, (LP) linear potentiometer /6.53/

6.8 Beam-Modulation Devices

In diffraction experiments which utilize fast detectors, beam chopping combined with
the use of phase-sensitive techniques can lead to a significant improvement in the
signal-to-noise ratio. In contrast to reactive scattering-surface techniques for
which frequency variation is an important asset /6.55/, a fixed frequency chosen
to lie in a low-noise regime is sufficient in diffraction experiments. In gas phase-
scattering experiments, modulation has often been carried out with a motor mounted
in the beam system, but other designs have been used which avoid the contamination
problems associated with this solution. Tuning-fork choppers /6.56,57/, piezoelec-
tric devices /6.58/, and magnetically-coupled feedthroughs /1.5/ have been incorporated into scattering system designs.

The beam-modulation device used in the authors' laboratory is shown in Fig.6.12
/6.59/. It consists of a rotary feedthrough flanged onto the system onto which a
chopper wheel is mounted from the inside. It is compatible with UHV environments
and can be rotated between 1,500 and 15,000 rpm. It is based on an asynchronous

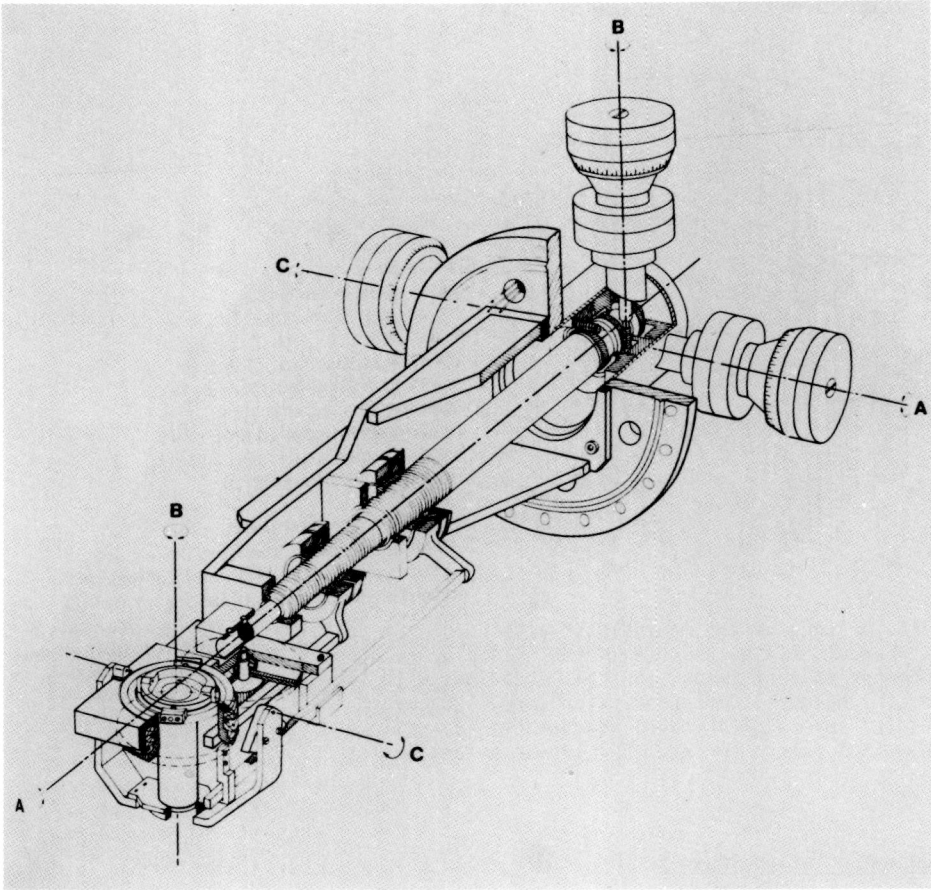

Fig.6.11. Schematic drawing of a UHV three-axis goniometer showing the rotation axes A, B, and C which interact at the surface /6.54/

motor; in this application the rotor is supported by bearings inside the vacuum system and the stator, which is more prone to outgassing, is mounted outside the vacuum envelope.

   Perhaps the most elegant modulation device is that based on a magnetic bearing which was developed at the Kernforschungsanlage Jülich /6.60/. Since there is no friction in operation, the rotation frequency is limited only by the mechanical properties of the rotor and chopper wheel. This and the rotary feedthrough discussed above are the most versatile devices, since they can be used for reactive scattering and time-of-flight investigations in addition to diffraction studies.

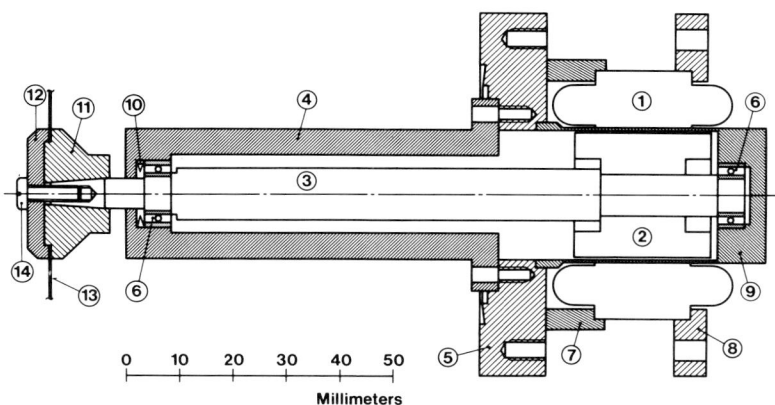

Fig.6.12. Schematic drawing of the beam-modulation device used in the authors' laboratory. (1) motor stator, (2) motor rotor, (3) shaft, (4) support cylinder, (5)
2 3/4" flange, (6) bearings, (7) and (8) stator supports, (9) vacuum envelope,
(10) spring washers, (11) and (12) chopper-disc clamps, (13) chopper disc, (14)
fastening screw /6.59/

6.9 Experimental Systems for Diffractive Scattering from Surfaces

The requirements for the individual parts of a gas-surface scattering apparatus
have been described in Sects.6.1-7. In this section we shall describe four different approaches to combine these parts to construct an apparatus for diffractive scattering.

Figure 6.13 shows the apparatus used by the Genoa group /3.20/ to study diffraction from a number of surfaces. The crystal surface S is mounted on a three-axis
goniometer and can be rotated in the scattering plane. The goniometer is in thermal contact with the liquid helium in cryostat II which allows substrate temperatures of 10 K to be reached. The detector is a liquid-helium cooled bolometer which
can be rotated in the scattering plane. Rotating cryostats I and II also act as
high-speed pumps for gases other than He and H_2 and generate a UHV environment in
the vicinity of the surface. The helium generated by the beam is pumped away by
oil-diffusion pumps mounted on the nozzle-skimmer and scattering chambers. The sensitivity of the apparatus is approximately 5×10^{-4} I_0.

A second apparatus which uses both oil-diffusion and cryopumping techniques is
shown in Fig.6.14 /6.43/. It consists of nozzle-skimmer (A) and collimator (B) chambers pumped by oil-diffusion pumps and the UHV scattering chamber (C) pumped by
liquid-helium cooled copper surfaces. Their surfaces were coated with a porous
Ca-Ag alloy which can be used to pump a limited amount of helium. The capacity is
sufficiently high that no other pumps are needed in the scattering chamber. The
crystal surface and the stagnation-detector orifice are located in the region la-

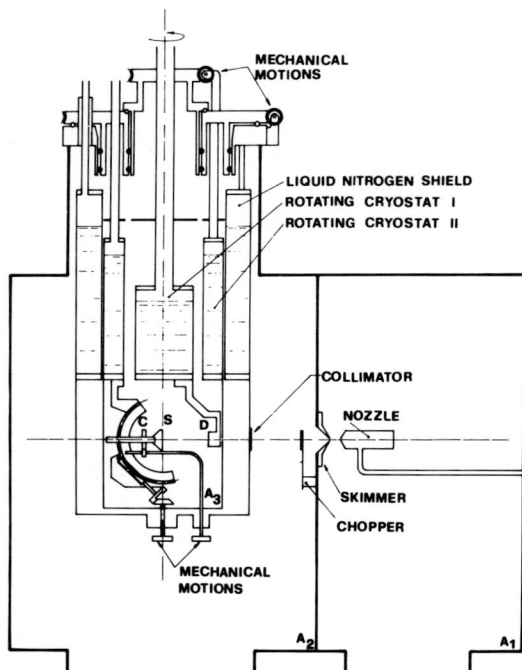

Fig.6.13. Schematic drawing of a cryogenic system with a bolometer detector for gas-surface scattering studies /3.20/

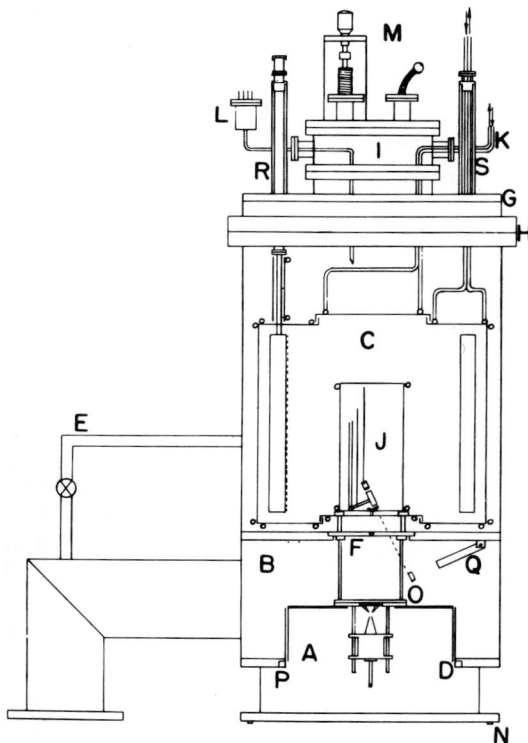

Fig.6.14. Schematic drawing of a cryogenic system with a stagnation detector for gas-surface scattering. (A) nozzle skimmer chamber, (B) differential pumping chamber, (C) scattering chamber, (L) detector. The substrate and detector orifice are in the region labeled J /6.43/

beled J. The ionization gauge detector L is mounted outside the low-temperature zone and is coupled to the scattering region with small diameter tubing and flexible stainless-steel bellows. The time constant is 30 s which limits the data accumulation rate. The sensitivity of the apparatus was not stated, but based on a later publication /6.61/, we estimate it to be between 10^{-3} and 10^{-4} I_0.

An apparatus which uses commercially-available UHV components is shown in Fig.6.15 /6.62/. It has the novel feature that the beam source is rotated within the scattering chamber, while the detector remains fixed. All other pumps are either of the liquid-nitrogen baffled oil-diffusion or titanium-sublimation type. Typical operating pressures during a He-beam scattering experiment are 10^{-3} Pa in the nozzle-skimmer chamber and 10^{-8} Pa in the detector chamber. Although the pumping speed in the nozzle-skimmer chamber is conductance limited to 500 l s^{-1}, a $P_0 d$ value of 10 can be attained for He. The sensitivity of the apparatus was not stated, but based on a later publication /6.63/, we estimate it to be 1×10^{-4} I_0.

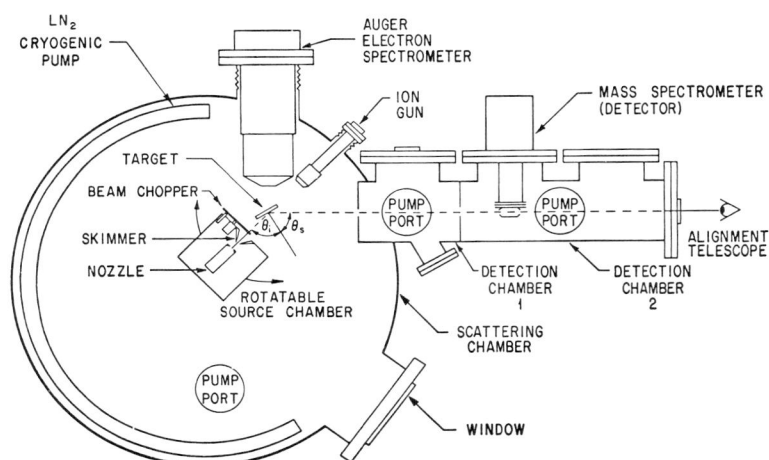

Fig.6.15. Schematic drawing of a molecular-beam surface-scattering apparatus in which the nozzle-skimmer assembly is rotated and the detector is fixed /6.62/

Figure 6.16 shows an apparatus which incorporates LEED and AES for surface analysis, and which utilizes a differentially-pumped ionization detector /6.64/. Chamber 1 contains the nozzle and skimmer whose arrangement can be seen in more detail at the lower left. This chamber is pumped with a water-baffled 500 l s^{-1} oil-diffusion pump. Chamber 2 contains the motor-driven chopper and is pumped with a liquid-nitrogen baffled 1250 l s^{-1} oil-diffusion pump. Chamber 3 is the UHV scattering chamber and is pumped with three 15 cm liquid-nitrogen baffled oil-diffusion pumps. The

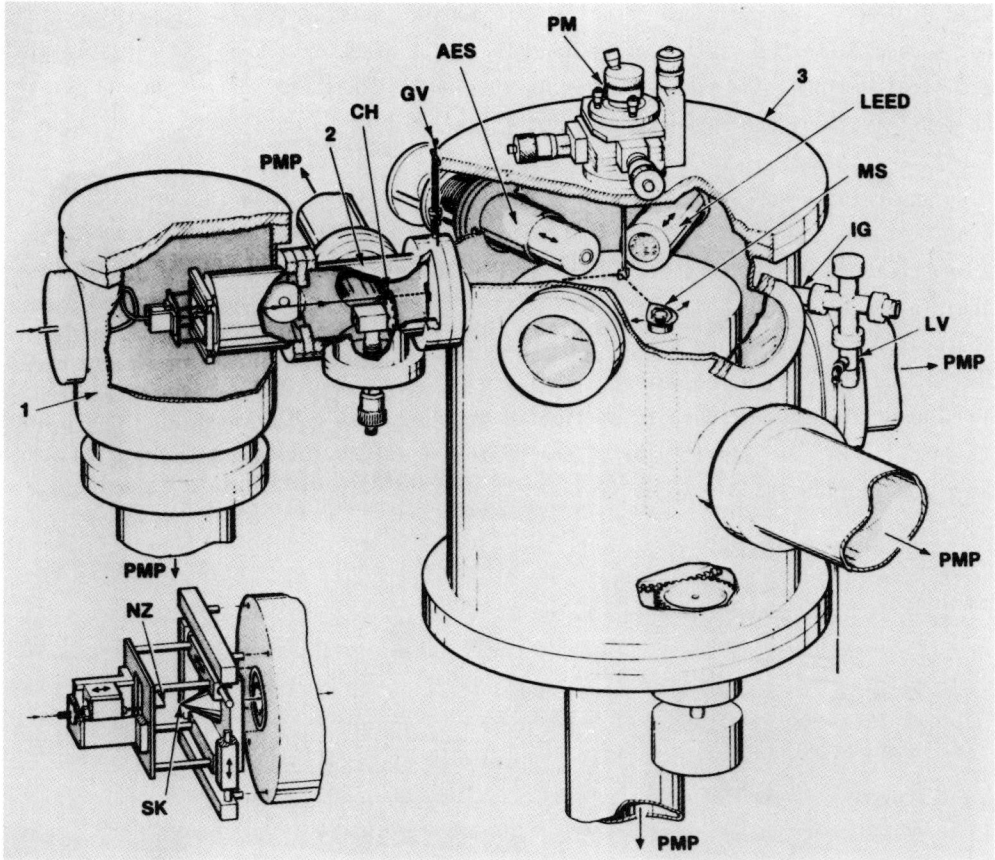

Fig.6.16. Schematic drawing of a molecular-beam surface-scattering apparatus. The three chambers are labeled 1, 2, and 3. At the lower left is an enlargement of the nozzle (NZ) and skimmer assembly (SK) contained in chamber 1. Chamber 2 contains a chopper (CH) mounted on a micrometer head. Between chambers 1 and 2 is a butterfly valve, and between chambers 2 and 3, a gate valve (GV) with three collimation apertures. Inside chamber 3 is an Auger spectrometer (AES) and LEED apparatus on bellows, a rotatable differentially pumped quadrupole mass spectrometer (MS), a precision manipulator (PM), and an ion gun (IG) /6.64/

quadrupole-mass-spectrometer detector is housed in the large cylinder visible in the scattering chamber. It rotates together with the housing, and the differential pumping is achieved by including a very small gap between the rotating cylinder and the stationary pump flange opening. The sensitivity of the apparatus is approximately 10^{-5} I_0 for long time constants /6.46/

   Each of these form designs has advantages for different types of investigation. If a cryopumped design is used, a fast bolometer detector /6.47/ would be a better choice than a stagnation detector, since modulated-beam and time-of-flight measurements are possible. The Genoa design is well suited for high-angular-resolution measurements for which a high sensitivity is required. Furthermore, cryogenic systems are ideally suited for low-temperature studies such as rare-gas adsorption on solids. The metal UHV systems are better suited for experiments on reactive surfaces for which LEED and AES are necessary. However, it is more difficult to work with liquid-helium cooled detectors and substrates in such systems because of the thermal radiation load.

6.10   The Influence of the Transfer Width of the Apparatus and of Surface Perfection on Measured Intensities

Both LEED and atom diffraction are techniques in which the diffraction pattern arises through the interference between the plane-wave components of the wave packet which represents each incoming particle. (For both methods, the incoming particle intensity is 10^{12}-10^{13} cm^{-2} s^{-1}, i.e. so low that interactions between particles can be neglected.) As the detectors in both techniques are not sufficiently sensitive to detect a single event, the experimental intensities are enhanced by the superposition of many individual diffraction events in a given time interval. Each wave packet gives rise to a diffraction pattern and the intensities of all these events are added to produce the final measured intensity. The question which arises in interpreting such diffraction data is to what extent the instrument itself and to what extent the crystalline perfection of the surface limit the widths of the diffracted wave packet(s) characterizing an individual diffraction event. Since the observed diffraction spots represent the superposed intensities of many such events, the widths in question are closely related to those of the measured spots as long as the latter do not overlap appreciably.

The problem of the instrumental limitation in LEED has been discussed by a number of authors /6.65-67/. PARK et al. /6.65/ has introduced the concept of a transfer width which represents the minimum lateral dimension over which the surface must be perfect to give diffraction spots whose widths are limited by instrumental resolution alone. It depends on the energy spread in the electron beam and on geometrical parameters of the instrument. For most LEED instruments, the transfer width is of the order of 100 Å /6.67/. COMSA /6.68/ has derived an explicit formula for the transfer width in atom diffraction experiments which again depends on the energy spread in the incoming beam and the geometrical parameters of the source and detector. The geometry of a typical experiment is shown in Fig.6.17. The angular spread at the detector due to the source and detector geometry is given approximately by /6.68/

138

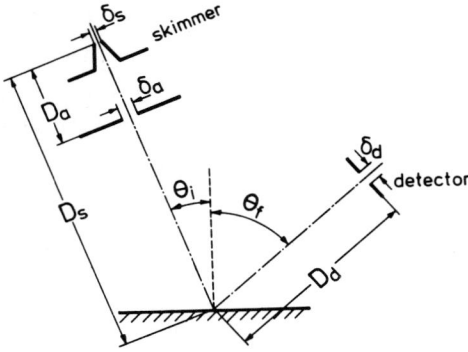

Fig.6.17. Schematic diagram of an apparatus for atom diffraction studies. δ_s, δ_a, and δ_d are the source, collimation aperture, and detector aperture diameters, respectively /6.68/

$$(\Delta_\theta \, \theta_f)^2 \sim \left(\frac{\cos\theta_i}{\cos\theta_f}\frac{\delta_s}{D_s}\right)^2 + \left(\frac{\cos\theta_i}{\cos\theta_f}\frac{\delta_a}{D_a}\right)^2 + \left(\frac{\cos\theta_f}{\cos\theta_i}\frac{D_s}{D_d}\frac{\delta_a}{D_a}\right)^2 + \left(\frac{\delta_d}{D_d}\right)^2 \quad , \tag{6.8}$$

and the contribution to the transfer width which arises from this geometrical factor, w_θ, is given by

$$w_\theta \sim \lambda/(|\Delta_\theta \, \theta_f|\cos\theta_f) \quad . \tag{6.9}$$

The energy-spread contribution to the transfer width arises from the fact that intensity maxima in the diffraction pattern from a particle of a given energy can overlap with intensity minima from a particle of different energy. This contribution to the transfer width can be written as

$$w_E \sim \lambda/\{|\sin\theta_i-\sin\theta_f|\left[\overline{(\Delta E)^2}/E^2\right]^{1/2}\} \tag{6.10}$$

where $\overline{(\Delta E)^2}$ is the mean-square energy spread in the beam. Combining the two vectorially, Comsa obtains

$$w \sim \lambda/\left[(\Delta_\theta \, \theta_f)^2\cos^2\theta_f + (\sin\theta_i-\sin\theta_f)^2 \, \overline{(\Delta E)^2}/E^2\right]^{1/2} \quad . \tag{6.11}$$

For $\theta_i = \theta_f$, corresponding to specular reflection, the energy dependence disappears and the transfer width when extracting information from the specular beam will be greatest as grazing angles are approached.

Large transfer widths can be obtained by minimizing the angular divergence in the beam and the beam diameter at the surface, as well as by using a small detector aperture and as large a surface to detector distance as possible. However, all of these steps will lead to a signal reduction at the detector; therefore, when designing a atom-diffraction system, there will be a tradeoff between transfer width and signal intensity. At present, it appears that using beam diameters and detector apertures of roughly 0.2 mm diameter transfer widths roughly equal to those in LEED can be obtained with He nozzle beams.

The above factors describe the limitations which the instrument imposes on the diffraction experiment. An additional angular broadening of the observed beams will be observed if the crystal is not well-ordered over dimensions which are small in comparison with the transfer width. This problem has been considered by LAPUJOULADE et al. /4.11/. They derived a formula for the case in which scattering occurs coherently over a lateral length ℓ_c and the intensities from all such areas on the surface add incoherently. For in-plane scattering they obtain the following expression for the angular distribution in the intensity I:

$$I \sim \ell_c \left(\frac{\sin[1/2\ell_c \Delta K]}{1/2\ell_c \Delta K} \right)^2 \quad , \tag{6.12}$$

where the in-plane tangential momentum change ΔK is given by

$$\Delta K = |k|(\sin\theta_i - \sin\theta_f) \quad . \tag{6.13}$$

This leads to a broadening in addition to the instrumental effects discussed above, and the total mean-square contributions from both effects can be written in the form

$$\overline{\Delta\theta^2_{exp}} = \overline{\Delta\theta^2_{instr}} + \overline{\Delta\theta^2_{fz}} \quad , \tag{6.14}$$

where the subscripts indicate the experimentally determined, instrument limited, and finite-size limited contributions to the angular widths, respectively. Since $\overline{\Delta\theta^2_{instr}}$ can be determined by measuring the angular distribution of the intensity in the direct beam, $\overline{\Delta\theta^2_{fz}}$ can be determined. LAPUJOULADE et al. /4.11/ have carried out such an analysis for He and Ne scattering from Cu(100) and, assuming that the specular peak broadening is due to purely elastic scattering, obtain a characteristic length of 78 Å.

The assumption that all the intensity in a measured scattered beam envelope is purely elastic is not always justified, although it should be valid for cases such as He scattering from metal surfaces at low temperatures where the specular intensity can be as high as 70% of the incoming intensity. However, in some cases a pronounced broadening of a diffracted beam is observed, as for H scattered from KCl(100) for which results are shown in Fig.6.18 /6.69/. This distribution can be separated into three parts: elastic, quasi-elastic, and inelastic plus incoherent elastic. The form of the elastic contribution is given by the direct beam profile; WILSCH et al. assumed that the quasi-elastic contribution has a Gaussian distribution and that the

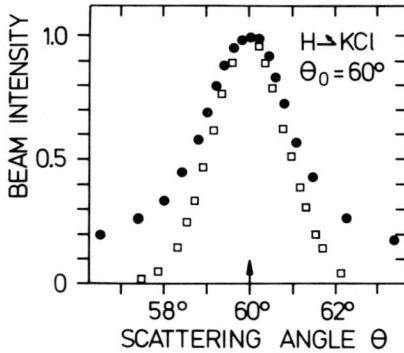

Fig.6.18. A comparison of beam profiles for H scattered from KCl(100). (□) denotes the direct beam and (●) the specular beam for θ_i = 60° /6.69/

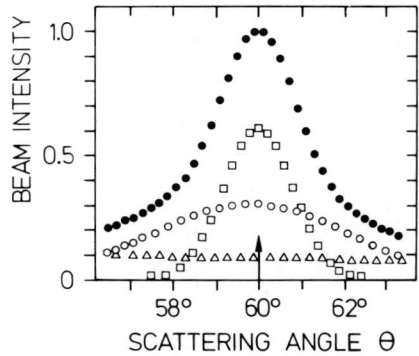

Fig.6.19. Separation of the specular beam shown in Fig.6.18 into cosine background (△), coherent elastic (●) and incoherent (o) intensities /6.69/

inelastic plus incoherent elastic background has a cosine distribution. With these assumptions, the individual contributions to the specular peak shown in Fig.6.18 are presented in Fig.6.19. Angular integration shows that the coherent elastic, quasi-elastic and cosine contributions represent 38.8%, 26.7%, and 34.5% of the total intensity, respectively. These results show that careful examination of the peak profiles is necessary to establish whether the raw data must be deconvoluted to obtain the elastic part before carrying out a structural calculation.

7. Structural Investigations on Surfaces of Ionic Crystals

7.1 Diffraction Studies on LiF(100)

The very first atom-diffraction studies were performed in 1929 with effusion sources by ESTERMANN and STERN /1.1/ on the system He/LiF(100) in order to establish the de Broglie relation (3.1). Since then this system has been investigated in great detail, and the first corrugation function from which structural implications were deduced has been reported for this system.

More than forty years after ESTERMANN and STERN, BOATO and coworkers /7.1,3.20/ published the first diffraction patterns obtained with He nozzle beams of good monochromaticity, and examples of their results with λ_i = 0.57 Å are shown in Fig.7.1 for two different in-plane scans corresponding to different azimuthal orientations

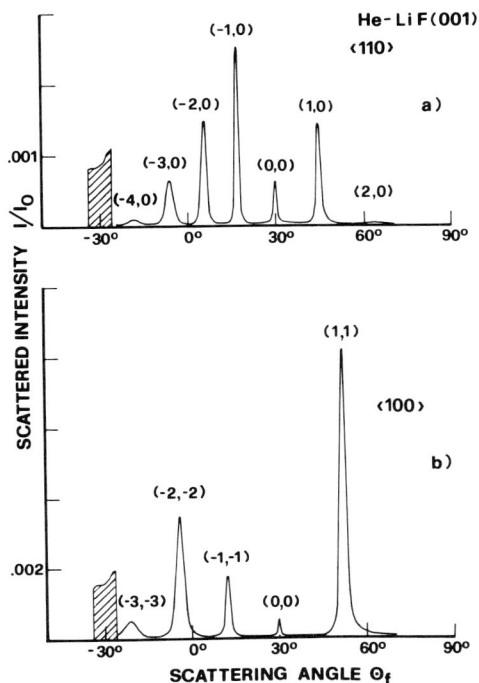

Fig.7.1a,b. He-diffraction spectra from LiF(100) at 80° K for θ_i = 30° with λ_i = 0.57 Å. Incident beam along (a) [110] azimuth, (b) [100] azimuth /3.19/

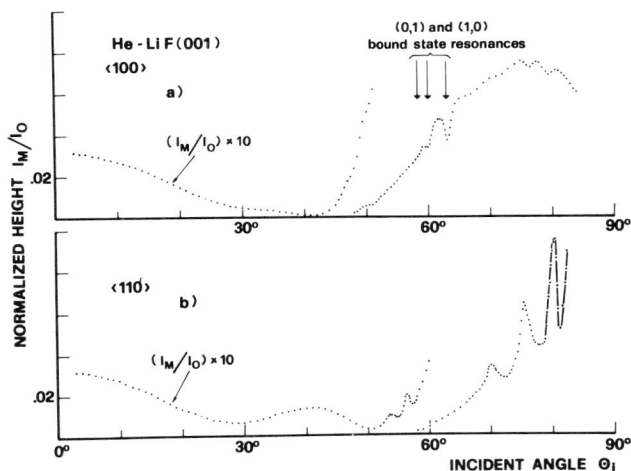

Fig.7.2a,b. Intensity of the specular beam as a function of θ_i for He/LiF (λ_i = 0.57 Å). Incident beam along (a) [110] azimuth, (b) [100] azimuth. Bound-state resonances play a role for θ_i > 50° /3.19/

of the sample. Figure 7.2 shows the intensity of the specular beam for these two orientations as a function of the angle of incidence. These curves exhibit a smooth variation until θ_i ~ 50° where selective adsorption features start to cause rapid angular intensity variations. These data constituted the basis for the first thorough structural surface analysis based on He diffraction, which was performed by GARCIA /3.17,18/. Using the GR method (Sect.3.4.1), he established the best-fit corrugation function by fitting the angular variation of the intensities of several beams for two azimuthal orientations as shown in Fig.7.3. The best-fit corrugation has the analytical form

$$\zeta(x,y) = \frac{1}{2}\,\zeta_{10}\left[\cos\frac{2\pi x}{a} + \cos\frac{2\pi y}{a}\right] + \frac{1}{2}\,\zeta_{11}\left[\cos 2\pi\,\frac{(x+y)}{a} + \cos 2\pi\,\frac{(x-y)}{a}\right] \tag{7.1}$$

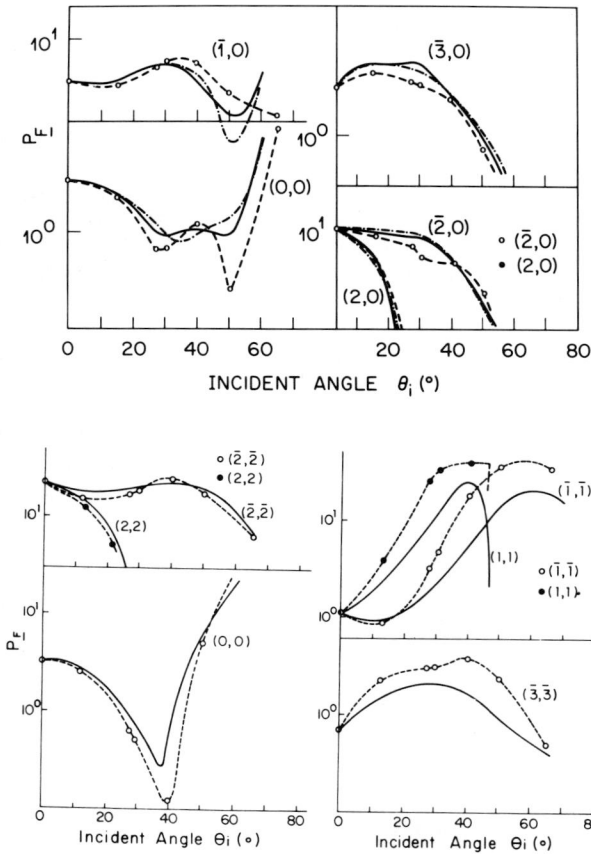

Fig.7.3. Logarithms of diffracted intensities versus angle of incidence θ_i for several beams. Points indicate experimental results for He/LiF obtained by BOATO et al. /3.19/. Full lines correspond to best-fit calculations of GARCIA /3.18/. Incident beam in upper part along [100] azimuth, in lower part along [110] azimuth. The dashed lines in the upper part correspond to calculations with the eikonal approximation using the best-fit corrugation of GARCIA /1.11/

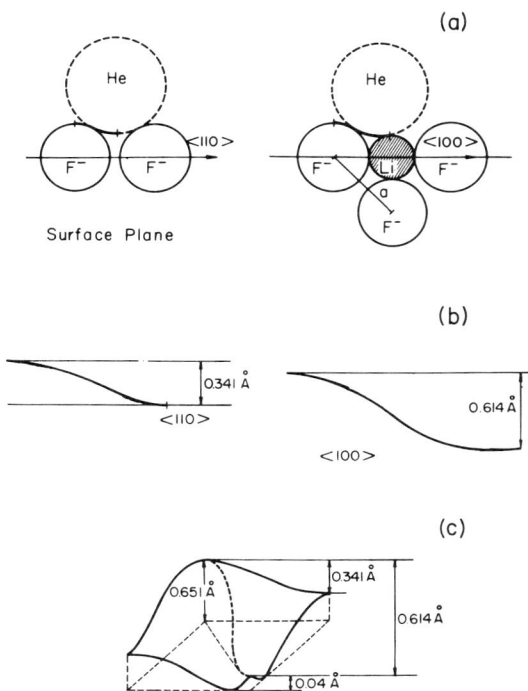

(a)

(b)

(c)

Fig.7.4. Geometrical representation
of the corrugation function of He/LiF
with the contact hard-sphere model
along the [100] and [110] azimuths
/3.18/

with ζ_{10} = 0.307 Å and ζ_{11} = 0.017 Å and the potential well depth was assumed zero.
Garcia proposed a very pictorial explanation for the shape of the corrugation as
shown in Fig.7.4: The corrugation (7.1) is reproduced very well by rolling the He
atom with radius r(He) = 1.78 Å (corresponding to spheres in contact in solid He)
over the LiF surface, whereby the points of contact of the He atom with the surface
ions (and not the up and down motion of the He centre) constitute the corrugation.

   The maximum amplitude of the corrugation function (7.1) is 0.62 Å which is practically identical to the difference of Pauling ionic radii r(F$^-$)-r(Li$^+$) = 1.33-0.68 =
0.65 Å /7.2/. This is a very reasonable result in view of the rigidity of these
two closed-shell ions, which establishes itself in their small polarizabilities or
deformabilities /7.3/. It indicates that, on the whole, the ionic bonding is pre-
served near and at the surface.

   It should be noted that BOATO et al. /3.20/ using the eikonal approximation, as
well as GARCIA /3.19/ using the GR method obtained a good fit for the [100] azimuth
with only one parameter, ζ_{10} = 0.301 Å. However, neglection of the coefficient ζ_{11}
yielded a worse fit for the [110] azimuth. Taking into account an (estimated) well
depth of 5 meV (Sect.5.1), BOATO et al. arrived at ζ_{10} = 0.289 Å. Although all these
values give a maximum corrugation amplitude ζ_m = 2ζ_{10} which is smaller than the dif-

ference in ionic radii, one should be cautious in interpreting this fact as being due to outward relaxation of the smaller Li^+ ion as proposed by GARCIA /3.17,18/. The different coordination number of the surface atoms certainly gives rise to different ionic radii at the surface. Furthermore, all theoretical investigations predict a small outward relaxation of an ion for the (100) surface of fcc alkali halides /7.4/. A LEED analysis of LiF(100) by LARAMORE and SWITENDICK /7.5/ which supported this conclusion, does not seem reliable, as the electron beam gives rise to a decomposition of the surface. Unfortunately, no energy dependence of the corrugation function or band splitting in resonant scattering has been studied up to now to establish the degree of rigidness of the closed-shell ions at the surface. However, the fact that the low-energy selective adsorption data of FRANKL et al. /5.11/ could be reproduced with the corrugation (7.1) by GARCIA et al. /5.24/ seems to indicate that the shape and maximum corrugation amplitude do not change very much between 0 and 70 meV.

In their theoretical study of He diffraction from LiF(100), GOODMAN and TAN /5.13/ describe the particle-surface interaction by a Morse potential (2.6) with the parameters D = 7.6 meV, σ = 1.1 $\overset{\circ}{A}^{-1}$, β_{10} = 0.10, and β_{11} = 0.007. An extrapolation to energies above zero has been performed by HOINKES /1.10/ and resulted in values for ζ_{10} of 0.265, 0.308 and 0.317 $\overset{\circ}{A}$, and for ζ_{11} of 0.0209, 0.0241, and 0.0248 $\overset{\circ}{A}$ corresponding to energies of 10, 50, and 90 meV, respectively. The result for 50 meV agrees very well with the result of GARCIA /3.18/ obtained for 63 meV. However, the dependence on energy is certainly too strong in view of the above discussion, so that one may conclude that the Morse potential does not describe the repulsive part of He/LiF properly (compare Sect.2.4).

BOATO et al. /7.1/ have also studied Ne diffraction from LiF(100). Typical diffraction spectra are shown in Fig.7.4. The background contains a great deal of structure as can be observed especially near the specular, and this structure has been correlated to scattering from surface phonons /7.6,7/ and coupling to bound states via surface phonons /7.8,9/. Both interpretations have resulted in attempts to establish the dispersion of the surface phonon Rayleigh branch /7.10/. Analysis of the elastic intensities based on (7.1) with ζ_{10} only and neglecting the potential well depths yielded ζ_{10} = 0.273 $\overset{\circ}{A}$. In view of Fig.7.4 this can be understood as being due to the increased radius of the Ne atoms compared to He.

Very recently, CARACCIOLO et al. /7.11/ have studied H_1 diffraction from LiF, and their analysis of the diffraction intensities based on (7.1) with the first term only yielded a value of ζ_{10} = 0.095 $\overset{\circ}{A}$. This is very small compared to the values derived from scattering with He and Ne and can hardly be attributed as being due to the larger H radius compared to Ne. A value of ζ_{10} = 0.16-0.20 $\overset{\circ}{A}$ as derived from the work of FINZEL et al. /2.2/ would fit more reasonably into the sequence of decreasing maximum corrugations with increasing radii of the scattered particles.

From their work on the temperature dependence of bound-state resonances in H/LiF, FRANK et al. /7.12/ came to the interesting conclusion that the lateral thermal expansion coefficient for large temperatures is about twice the value of the bulk.

Diffraction of H_2 from LiF(100) was also reported by BOATO et al. /7.1/. Figure 7.5 shows typical diffraction spectra for $\lambda = 0.76$ Å. Besides the diffraction peaks expected for this wavelenght, additional peaks with smaller intensity are found which correspond to rotational transitions of the H_2 molecules during scattering. In the scattering process, there is energy exchange between translational and rotational motion, but no energy exchange with the surface. The conditions for diffraction are then [in analogy to (3.8,9)]

$$\underline{K}_f = \underline{K}_i + \underline{G} \tag{7.2}$$

and

$$k_f^2 = k_i^2 \pm \frac{2m}{h^2} E_{rot} \quad , \tag{7.3}$$

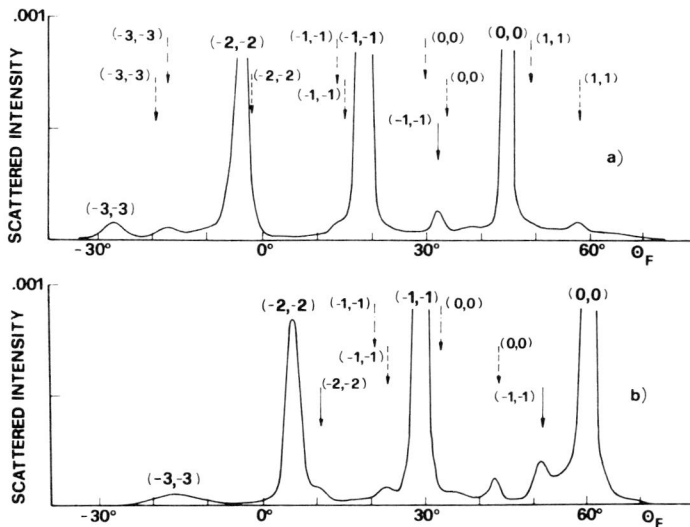

Fig.7.5. Diffraction pattern of H_2/LiF(100) with $\lambda_i = 0.76$ Å for (a) $\theta_i = 60°$, (b) $\theta_i = 45°$. Incident beam along [100] azimuth. Full, dashed, and dash-dotted arrows indicate the location of peaks corresponding to $(0 \to 2)$, $(2 \to 0)$, and $(3 \to 1)$ rotational inelastic transitions of H_2, respectively /7.1/

from which the angular location of the additional peaks is easily derived. For a one-dimensional corrugation with $G = j2\pi/a$, we obtain

$$\sin\theta_i = \frac{\lambda_f}{\lambda_i}\left(\sin\theta_i + j\frac{\lambda}{a}\right) \quad , \tag{7.4}$$

whereby $\lambda_f = 2\pi/k_f$ can be obtained from (7.3). Equation (7.4) yields, of course, (3.12) for $E_{rot} = 0$. In analogy to (7.4), for two-dimensional corrugations $\sin\phi_{j\ell}$ and $\sin\theta_{j\ell}$ can be obtained for rotational transitions by multiplying the right-hand sides of (3.13a,b) by λ_f/λ_i.

The energy change E_{rot} between neighbouring H_2 rotational states is given by $2(2L-1)k_B T_r$ (L > 2) with L being the rotational quantum number, k_B Boltzmann's constant, and T_r ($\cong$ 85 K) the characteristic rotation temperature of H_2. The peaks observed in Fig.7.5 correspond to transitions $0 \to 2$, $2 \to 0$, and $3 \to 1$. The transition $1 \to 3$ is not possible in this experiment, since the beam energy is smaller than the rotational-energy change $E_{rot}(1 \to 3) = 10\ k_B T_r$.

An analysis of the corrugation function was not performed for H_2/LiF, because of the appreciable rotationally inelastic contributions and because of the enhanced inelastic effects due to phonon excitations (compared to He diffraction). The latter can be observed from the smaller fraction of elastically scattered particles and by the large tails around the diffraction peaks. The reason for this enhanced inelastic scattering may be due to the fact that although the mass of H_2 is half that of He, its effective incident energy $E_i' = E_i + D$ with $D \cong 40$ meV is about doubled. A rough estimate of the corrugation amplitude could be performed by using the rainbow angles (3.52) and yields $\zeta_{10} \cong 0.2$ Å, which is not unreasonable in view of the He and Ne results.

7.2 Diffraction Studies on NiO(100)

Another substance with highly ionic bonding character, whose (100) surface has been investigated in detail with particle beam diffraction, is NiO. Both scattering of He and H_2 were reported by CANTINI et al. /2.10/. The data for He diffraction were carefully analyzed later by using Patterson series /2.11/, and the best agreement between experiment and calculation was obtained with the Fourier coefficients $\zeta_{10} = 0.139$ Å, $\zeta_{11} = 0.009$ Å, $\zeta_{20} = 0.007$ Å, and $\zeta_{21} = 0.010$ Å, where (3.2) for a quadratic lattice with inversion symmetry was used. A contour map of this corrugation function is shown in Fig.7.6. The maximum corrugation amplitude is $\zeta_m = 0.28$ Å, which is much smaller than the difference of the bulk ionic radii of O^{2-} and Ni^{2+}: $r(O^{2-}) - r(Ni^{2+}) = 1.4 - 0.72 = 0.68$ Å /7.2/. LEED calculations /7.13,14/ fitting experimental inten-

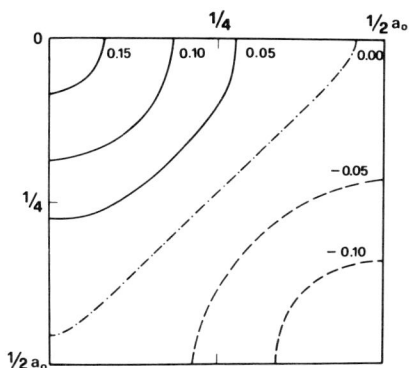

Fig.7.6. Contour map of the corrugation function for NiO(100) as obtained by CANTINI et al. /3.27/. Contour levels are given in Angström units. The maximum corrugation amplitude is 0.28 Å

sity versus voltage curves /7.15/ indicate that the location of the surface ion cores does not differ from the corresponding bulk locations by more than ±0.05 Å. Also, theoretical model calculations for surface relaxations of oxide ions predict only slight changes of the core locations /7.16/. The small corrugation amplitude obtained with He diffraction suggests therefore that at the surface of NiO(100) an appreciable charge redistribution must take place, the location of the cores being very nearly the same as in the bulk.

Diffraction of H_2 from NiO(100) again exhibits additional peaks due to rotational transitions and also shows appreciably stronger inelastic contributions due to phonon scattering than He diffraction. In analogy to H_2/LiF, the latter effect is probably again due to the larger potential depth for H_2/NiO (D ~ 60 meV) than for He/NiO (D ~ 20 meV). From the positions of the rainbow maxima, CANTINI et al. estimated the parameter ζ_{10} ~ 0.25 Å. It is remarkable, but not at all understood, that the NiO(100) surface is flatter than LiF(100) when seen by He, while it is more corrugated and probably looks more complex when seen by H_2 molecules /2.10/.

7.3 Diffraction from Other Ionic Materials

Rather early He-diffraction studies with nozzle beams from NaCl(100) and LiF(100) were reported by BLEDSOE and FISHER /7.17/. Since at low sample temperatures, contaminations of the surfaces (especially of NaCl) due to the poor background pressure of 10^{-7} Torr were obtained, the experiments were performed with the samples at room temperature. Whereas the diffraction patterns of LiF show rather small inelastic contributions, the scattering from NaCl is dominated by inelastic events due to the lower Debye temperature. Nevertheless, GARCIA et al. /7.18/ attempted to estimate the corrugation amplitude and found ζ_{10} = 0.34 Å. This amounts to a

maximum corrugation amplitude $\zeta_m \sim 0.7$ Å which is reasonably close to the difference of the bulk ionic radii of Cl$^-$ and Na$^+$: $r(Cl^-) - r(Na^+) = 1.81 - 0.98 = 0.83$ Å
/7.2/.

H$_1$ diffraction from KCl(100) was investigated by FRANK et al. /2.4/. For sample
temperatures ≤ 170 K, the small residual pressure of water gave rise to an ordered
water overlayer. A qualitative intensity analysis yielded $\zeta_{10} = 0.76$ Å which is very
large and lies outside the validity range of the eikonal approximation applied. However, in view of the one water molecule adsorbed per unit cell of KCl(100), this result does not seem unreasonable and a refined analysis using a large amount of diffraction data seems very worthwhile.

We close this section by mentioning the experiments of ROWE and EHRLICH /7.19,20/
who scattered He, H$_2$, HD, and D$_2$ from MgO(100). These authors used an effusion source
and were mainly interested in the rotational transitions during scattering of H$_2$, HD,
and D$_2$. For He diffraction, only beams of first order were well resolved and no information on the corrugation function could be obtained. This substance would be a
favourable candidate for more detailed studies, as reliable LEED analyses exist
/7.21-23/.

8. Structural Investigations on Semiconductor Surfaces

8.1 Helium-Diffraction Studies on Si(111) and Si(100)

Semiconductor surfaces show a more extensive reconstruction than metal surfaces due
to the directional nature of the dangling bonds left when the surface is exposed.
Furthermore, the reconstruction will extend deeper into the solid, and for Si(100)
is believed to extend into the fifth atomic layer /8.1/. In view of the technolo-
gical relevance of these surfaces and the large changes in the electronic properties which accompany small structural modifications /8.2/, structural determinations on semiconductor surfaces is an important research area. However, to date,
not much progress has been made in this field using LEED due to the complexity of
the reconstructions which occur. One example in which the structure has been solved
with LEED is GaAs(110) /8.3,4/, but neither the Si(111) nor the Si(100) surface
phases are structurally understood. Atom diffraction could be of use here, since
the scattering mechanism is simpler than in LEED and multiple-scattering effects
will not be as pronounced. However, the reconstructed semiconductor surfaces will
have a more open structure than metal surfaces, and therefore the incoming atom

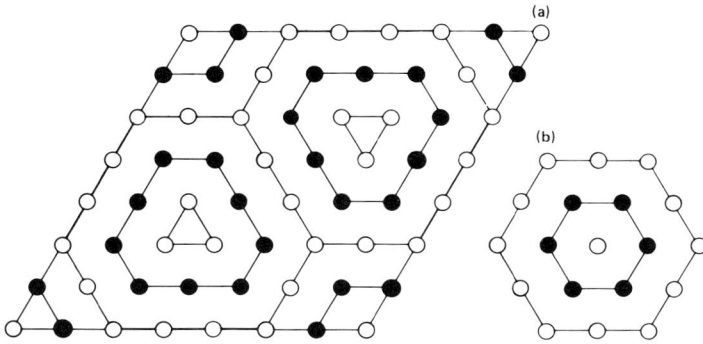

Fig.8.1

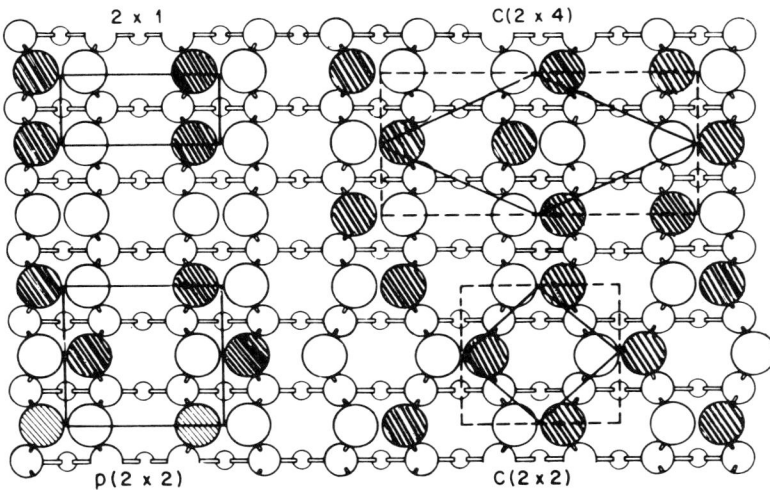

Fig.8.2

Fig.8.1. (a) Model proposed for the (7×7) surface of Si(111). (b) The hexagonal arrangement of the surface atoms about a corner atom of the unit cell is shown. Raised and lowered atoms are represented by open and shaded circles, respectively /8.5/

Fig.8.2. Various periodicities for the Si(100) surface based on the tilted dimer model of CHADI /8.2/. The shaded atom is raised with respect to its dimer partner. The alternation of the tilt will yield p(2×2) or c(2×2) periodicities depending on the phasing of adjacent rows /8.6/

will, in general, see more than the topmost layer of the surface. This will lead to a corrugation function which is strongly structured.

Before discussing recent atom-diffraction studies on silicon surfaces, we show models which have been proposed for the Si(111) (7×7) and the reconstructed Si(100) surfaces in Figs.8.1,2. Figure 8.1 shows the buckled ring-like structure proposed by CHADI /8.5/ for Si(111) (7×7) in which the open and shaded circles represent raised and lowered atoms, respectively. The model is consistent with the six-fold rotational symmetry seen with LEED at low energies at which diffraction from the topmost layer should dominate. Models for the Si(100) surface based on a tilted dimer model proposed by CHADI /8.2/ are shown in Fig.8.2. Depending on the surface preparation and annealing techniques, both (2×1) and c(2×4) periodicities have been reported /8.6/. The dominant feature of the reconstruction is the pairing of Si rows to form tilted dimers, which leads to doubling of the periodicity. The phases of the dimer tilts between adjacent paired rows allow (2×1), p(2×2), c(2×4), and c(2×2) structures to be formed.

He-diffraction studies on Si(100) by CARDILLO and BECKER /8.6/ show that several of the structures in Fig.8.2 are present together on the surface. Diffraction traces for various cuts through the reciprocal lattice are shown in Fig.8.3. Dominant peaks are seen for the integral and half-integral beams as expected from two perpendicular domains of a (2×1) structure. However, additional diffraction inten-

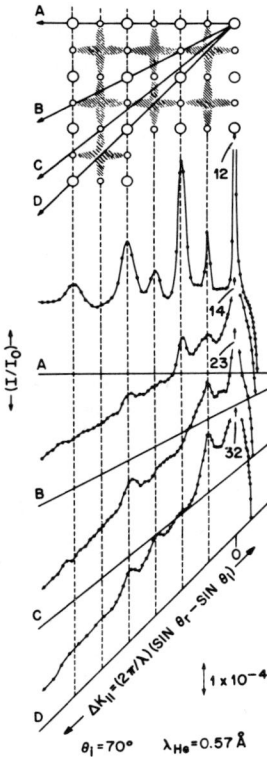

Fig.8.3. He-diffraction traces from Si(100) for four different azimuthal angles labeled A through D. The intensities normalized to the incoming intensity are plotted against the parallel momentum change ΔK, and the abscissa is rotated by the azimuthal angle ϕ. This allows a vertical correspondence between the reciprocal lattice and the diffraction trace. $\theta_i = 70°$, $\lambda_i = 0.57$ Å /8.6/

sity is seen in the regions of the reciprocal lattice shown shaded in Fig.8.3. This
diffuse region was also observed with LEED at low energies. The streaked pattern can
be interpreted as small domains of (2×2) and (2×4) structures which are longer, parallel to the row pairing direction than perpendicular to it. If the phases of adjacent domains are largely uncorrelated, streaking such as that seen in Fig.8.3 will
be observed. The lack of long-range order is consistent with the observation that in-
plane diffraction traces showed only 1-2% of the incoming beam intensity.

A diffraction trace for the Si(7×7) surface obtained by CARDILLO and BECKER /8.7/
is shown in Fig.8.4. The individual peaks can be clearly resolved for λ_i = 1 Å, and
a marked oscillation in the beam intensities is observed. Again, the in-plane scattering is only 1-2% of the incoming beam intensity, but considerable out-of-plane
intensity is seen using a larger detector aperture. This suggests that 10-20% of
the incoming intensity appears in the diffraction peaks. For both these surfaces,
which are expected to show a large corrugation, numerous diffraction peaks are observed whose intensities vary rapidly with the wavelength and angle of incidence.
However, at present no model consistent with the intensities observed has been reported. The added complexity of incomplete long-range order together with a strong
two-dimensional corrugation make these surfaces less accessible to a straightforward intensity analysis than weakly corrugated surfaces. However, with advances in
our understanding of the gas-surface interaction potential and in the calculation
of corrugation functions from measured diffraction intensities, it is to be expected that these surfaces can be analyzed.

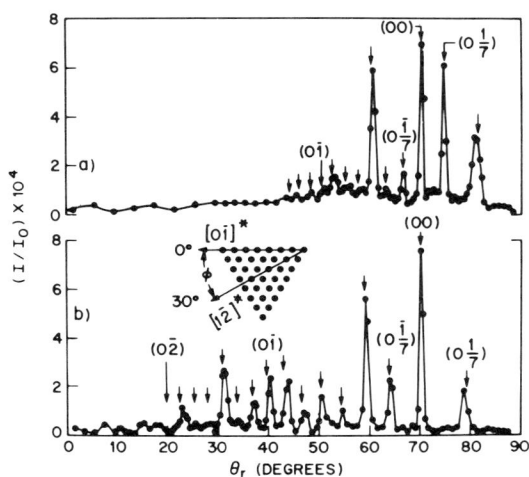

Fig.8.4. He-diffraction trace on the (7×7) structure of Si(111) for (a) λ_i = 0.57 Å,
and (b) λ_i = 1.0 Å. The expected angular positions of the seventh-order beams are indicated by arrows. The inset shows the reciprocal lattice. θ_i = 70°, ϕ = 0° /8.7/

8.2   Helium Diffraction from GaAs(110)

The complexity of the scattering data for silicon surfaces suggests that atom diffraction should first be carried out on a semiconductor surface of known structure.
This has been done recently by CARDILLO et al. /8.8/ who have investigated He diffraction from GaAs(110). The structure of this surface, which has been established
with dynamical LEED calculations /8.3,4/ is shown in Fig.8.5. The side view illus-

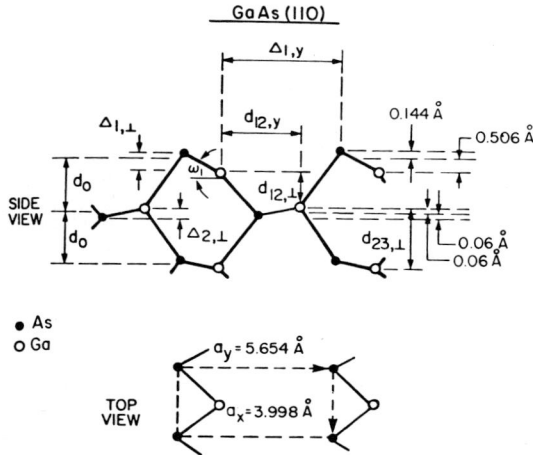

Fig.8.5. Schematic diagram of the geometry of the GaAs(110) surface /8.3/

trates that the surface has deep troughs parallel to the $[1\bar{1}0]$ direction. It will,
therefore, be more strongly corrugated in the $[001]$ than in the $[1\bar{1}0]$ azimuth. Furthermore, since the Ga and As atoms in the topmost layer are not coplanar, the surface is not symmetrical when viewed from opposite directions in the $[001]$ azimuth.
Figure 8.6 shows He-diffraction traces in this azimuth, where $\phi = 0°$ and $\phi = 180°$
correspond to an atom incident from the left and the right, respectively, in Fig.
8.5. The asymmetry can best be seen for angles nearest grazing and is less pronounced for $\theta_i = 35°$. Strong oscillations of the specular-beam intensity with θ_i
are observed which are most pronounced for long wavelengths as is shown in Fig.8.7
for the beam incident in the $[1\bar{1}0]$ azimuth. This is indicative of resonant scattering from the bound states in the gas-surface potential and complicates the structural analysis. Cardillo et al. have attempted to fit their data for $\lambda = 0.98$ Å to
a hard-wall model using the eikonal approximation and a corrugation function of the
form

$$\zeta(\underline{R}) = \frac{\zeta_1}{2} \cos \frac{2\pi x}{a_1} + \frac{\zeta_2}{2} \cos \frac{2\pi y}{a_2} \qquad (8.1)$$

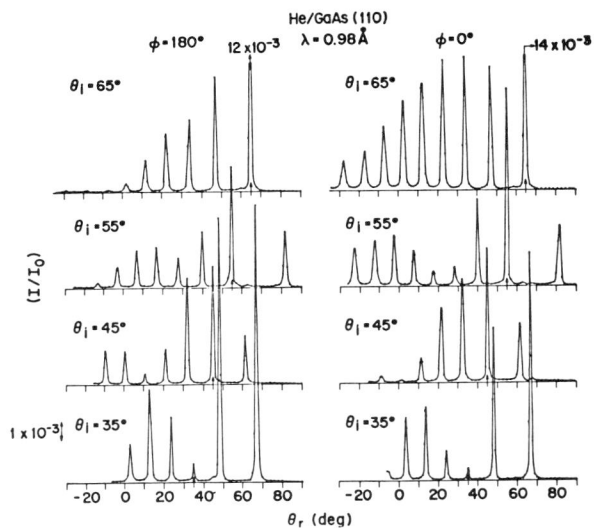

Fig.8.6. He-diffraction traces from GaAs(110) for various angles of incidence θ_i and for two azimuthal angles (see text). λ_i = 0.98 Å /8.8/

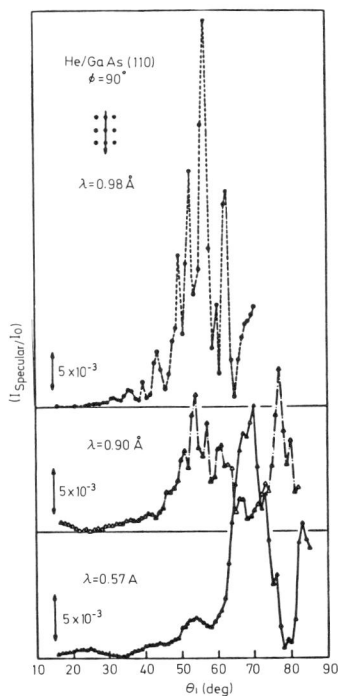

◄ Fig.8.7. Normalized specular intensity for He scattering from GaAs(110) as a function of θ_i for three wavelengths λ_i /8.8/

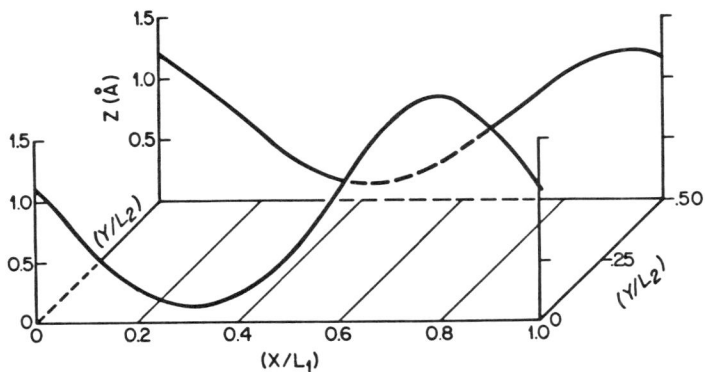

Fig.8.8. The corrugation function for GaAs(110) given by (8.1) with the parameters determined in /8.8/

with a_1 = 5.65 Å and a_2 = 3.998 Å. A degree of qualitative agreement is obtained
for ζ_1 = 1.2 Å and ζ_2 = 0.36 Å, and the corrugation function given by (8.1) is
shown in Fig.8.8. This data analysis is strongly simplified in that it does not
include the asymmetry of the surface in the [100] azimuth, that it neglects resonant scattering which can significantly alter the relative beam intensities at a
given angle of incidence for the wavelength used in the analysis, and that the
eikonal method is not valid for such large corrugations. A more detailed analysis
can only be carried out with a theoretical model which includes multiple-scattering and the attractive potential at the surface.

8.3 Diffraction from Graphite

The basal plane of graphite [the (0001) surface] has been extensively studied by a
number of investigators. The main emphasis of this work has been on the attractive
part of the atom-surface potential and the results have been discussed in Chap.2.
Here, we shall review the structural aspects of these studies. Helium-diffraction
data of BOATO et al. /2.36/ for λ = 0.57 Å taken at a substrate temperature of 80 K
were fit to a hard-wall model using the corrugation function

$$\zeta(\underline{R}) = \frac{1}{2}\zeta_{10}\left(\cos\frac{2\pi x}{a} + \cos\frac{2\pi y}{a} + \cos\frac{2\pi(x-y)}{a}\right) \tag{8.2}$$

which is consistent with the hexagonal open structure of this surface and in which
a = 2.465 Å is the hexagonal unit-cell length. A good fit to the data is obtained
with (8.2) for ζ_{10} = 0.092 ± 0.008 Å which corresponds to a maximum corrugation amplitude of 0.21 Å.

As was shown in Chap.2, the gas-surface potential can be determined by analyzing
resonant-scattering data. From the lateral variation of the repulsive part of the
potential, an energy-dependent maximum corrugation amplitude can be obtained. CARLOS
and COLE /2.35/ obtain values for ζ_{10} of 0.116 Å and 0.128 Å corresponding to equi-
potentials of 0 and 60 meV, respectively. This is in reasonable agreement with the
hard-wall corrugation amplitude obtained independently from analyses of the diffraction intensities (Sect.2.3). Equipotential surfaces calculated by summing pair-wise
potentials by STEELE /8.9/ result in a value of ζ_{10} which is considerably smaller
than the experimental value, for reasons which are at present not clear. However,
these calculations show that all higher Fourier coefficients have negligible amplitudes, which supports the one-parameter corrugation function of (8.2).

8.4 Diffraction from Layer Compounds

Helium-diffraction has also been applied to the study of charge-density waves on
TaS_2. This material is composed of layers, each of which can be thought of as an
array of Ta ions sandwiched between sheets of sulfur ions. The layers are weakly
bound to one another through van der Waals's forces. This compound undergoes a reconstruction below 80 K driven by a charge-density wave. This results in a modulation of the local charge, and a rearrangement of the Ta ions, which produce a deformation of the sulfur array both by locally changing the ionic radius of the tantalum and by a reaction of the sulfur to the tantalum movement /8.10/.

   Helium-diffraction studies by CANTINI et al. /8.10,11/ have shown that the surface has the same charge-density wave superlattice as the bulk. However, the superlattice peaks are of approximately the same intensity as those corresponding to the
unreconstructed structure, whereas they are a factor 20-50 weaker in x-ray, neutron,
or electron scattering. This indicates a large periodic deformation at the surface
of TaS_2 and shows the sensitivity of atom diffraction to charge-density wave structures. Figure 8.9 illustrates the reciprocal lattice of the unreconstructed lattice
and that for the $\sqrt{13} \times \sqrt{13}$ R 13°54' phase. Figure 8.10 shows diffraction traces in the
azimuth of the charge-density wave peaks and in the azimuth which would correspond
to the unreconstructed structure. All of the peaks expected are found, and in addition, other peaks, such as that between the (00) and ($\bar{1}$0) beams in Fig.8.10a, are
observed. The origin of these extra peaks is not known.

   The simplest corrugation function which will describe $\sqrt{13} \times \sqrt{13}$ R 13°54' periodicity contains the terms

$$\zeta_0(\underline{R}) = \frac{1}{2}\zeta_0 \left[\cos\frac{2\pi}{a_0\sqrt{13}}(4x-y) + \cos\frac{2\pi}{a_0\sqrt{13}}(x+3y) + \cos\frac{2\pi}{a_0\sqrt{13}}(3x-4y) \right] \qquad (8.3)$$

and

$$\zeta_{cdw}(\underline{R}) = \frac{1}{2}\zeta_1 \left[\cos\frac{2\pi}{a_0\sqrt{13}}x + \cos\frac{2\pi}{a_0\sqrt{13}}y + \cos\frac{2\pi}{a_0\sqrt{13}}(x-y) \right] \qquad , \qquad (8.4)$$

where x and y are the coordinates in the reconstructed surface lattice and $a_0 = 3.38$ Å
is the basic lattice parameter.

   The first of these formulas describes the unreconstructed surface, and the second
reflects the modulation induced by the charge-density wave. Reasonable agreement with
experiment is attained using the eikonal approximation for $\zeta_0 = 0.22 \pm 0.04$ Å and
$\zeta_1 = 0.16 \pm 0.04$ Å. These values give a maximum corrugation amplitude of 0.28 Å. The
authors interpret the corrugation amplitude as the displacement of the sulfur atoms

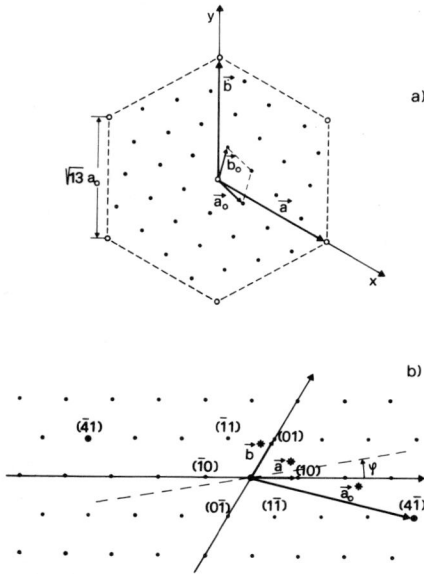

Fig.8.9. (a) Direct, and (b) reciprocal lattice of the $\sqrt{13}\times\sqrt{13}$ R 13°54' phase of TaS$_2$ /8.10/

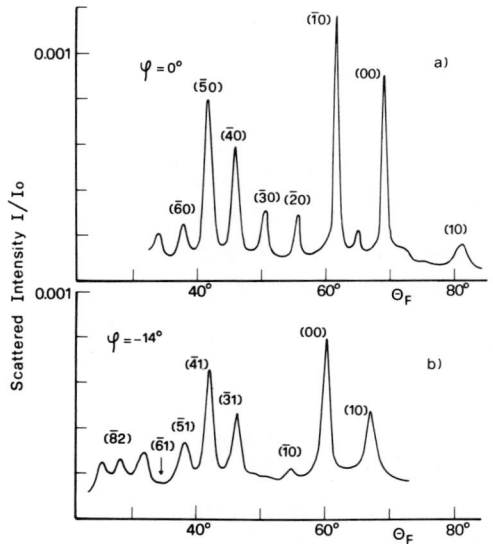

Fig.8.10a,b. He-diffraction traces from TaS$_2$. (a) This azimuth contains the unit-cell vector of the charge-density wave superlattice θ_i = 69°. (b) This azimuth contains the unit-cell vector of the unreconstructed surface. θ_i = 60°. λ_i = 0.57 Å for both traces /8.10/

due to the changed ionic radius of the Ta ions. It is, therefore, the reaction of the sulfur to the charge-density wave rather than a measure of the charge-density wave itself.

8.5 Helium-Diffraction Studies from other Surfaces

Helium diffraction has recently been observed from an ordered silicide on a Pt(100) surface formed by segregation of a silicon impurity at the surface /8.12/. The maximum corrugation amplitude was estimated to be 0.07 Å from rainbow angles. A surface carbide leading to a (3×5) superstructure on W(110) was investigated with He and Ne scattering. Resolved diffraction features were seen with He, but not with Ne /8.13/.

9. Structural Investigations on Metal Surfaces

9.1 Introduction

Metal surfaces are expected to be less corrugated than those of semiconductor or
ionic crystals, since the metallic valence electrons are partially delocalized and
will smooth out much of the corrugation expected on the basis of a hard-sphere surface model. The degree to which the electrons smooth out surface roughness has been
examined theoretically /9.1/, and the dependence of the corrugation amplitude on
the atomic spacing in the surface can now also be determined theoretically for
simple metals /2.32/. Therefore, experimental determination of the corrugation at
metal surfaces can provide a valuable test for these theories. A further interesting area for research is that of surfaces with defects. Regularly spaced defects such as steps can give rise to spatial variations of the electron-charge
density which may play an important role in surface reactions /9.2/. Due to its
unique sensitivity to the electron-charge distribution, atom diffraction can be
used to investigate these surfaces. In the following, we shall review results obtained with helium diffraction from close-packed metal surfaces, from the less
densely packed structurally similar bcc(112) and fcc(110) surfaces, and from stepped
metal surfaces.

9.2 Helium and Hydrogen Diffraction from Close-Packed Metal Surfaces

On close-packed metal surfaces, the valence electrons smooth out the roughness expected on the basis of a hard-sphere model almost completely. Figure 9.1 shows a
diffraction trace for the scattering from Ag(111) /9.3/ taken with a high sensitivity. It is seen that the intensity of the first-order diffraction peaks observed
is only 3×10^{-3} of that for the specular peak. This indicates that the maximum corrugation amplitude is less than 0.01 Å. Boato et al. also observed H_2 diffraction
from Ag(111), and Fig.9.2 shows the intensity of the first-order diffraction peaks
normalized to that of the specular peak as a function of q_z^2, the square of the perpendicular momentum transfer where $q_z = k_i(\cos\theta_i + \cos\theta_f)$, for both He and H_2 diffraction. It is seen that the diffracted peaks are an order of magnitude more intense for H_2 than for He. This effect has also been observed in a more recent investigation of the same diffraction system /6.63/ and suggests that the potentials
seen by the incoming He and H_2 are sufficiently different that the surface appears
to have a different corrugation in the two cases. A recent theoretical investiga-

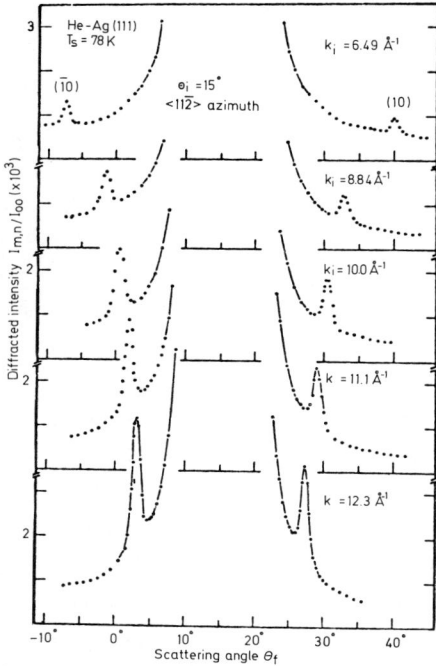

Fig.9.1. Diffracted intensity relative to the specular intensity as a function of the scattering angle θ_f for He diffraction from Ag(111). $\theta_i = 15°$, $k_i = 6.49$ Å^{-1}. The surface temperature T_s is 78 K /9.2/

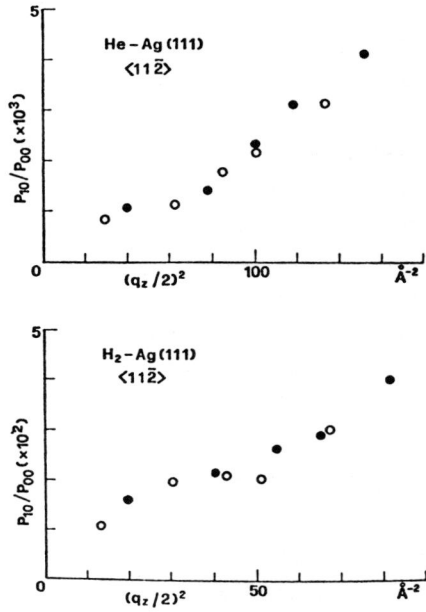

Fig.9.2. Diffracted intensity in the first-order peaks, for incident He and H_2 beams normalized to the specular intensity as a fuction of the square of the perpendicular transfer q_z^2. (o) designates (10) beams;(•) designates ($\bar{1}$0) beams /9.2/

tion of these results came to the conclusion that a corrugation in the attractive part of the potential for the more polarizable H_2 molecule could lead to the effect observed /9.4/. This example shows that the atom-surface potential must be more accurately known to extract absolute values for the surface corrugation from diffraction intensities.

He and H_2 (D_2) scattering has also been reported from Cu(100) /9.5/ and Pt(111) /9.6/. In both these cases, no features other than the specular peak were seen. Additional diffraction peaks have been reported in another study of Cu(100), but were attributed to surface impurities /4.17/. At present, the experimental evidence to data indicates that close-packed or nearly close-packed metal surfaces have corrugation amplitudes of less than 0.01 Å.

9.3 Helium Diffraction from fcc(110) and bcc(112) Planes

Both these planes consist of close-packed rows separated by an appreciably larger distance than the atom spacing in the rows. Generalizing from the results presented in Sect.9.2, one would expect the corrugation parallel to the rows to be negligible relative to that perpendicular to the rows. In fact, this has been observed for He diffraction from W(112) /9.7/ and Ni(110) /9.8/, so that viewed with atom diffraction, these surfaces appear to be one-dimensional.

Helium diffraction was first observed from a metal surface by TENDULKAR and STICK-NEY on W(112) /9.7/, and a later fit of the data to a hard-wall model by GOODMAN /9.9/ yielded a value of 0.16 Å for the maximum corrugation amplitude. This diffraction system has been recently reinvestigated /9.10/. Figure 9.3 shows more recent He-diffraction results obtained on a Ni(110) surface. The intensity of the (0 ±1) beams relative to the specular intensity is plotted as a function of the incident wavelength, and the data is fitted with a hard-wall model using the eikonal approximation. The best-fit corrugation function is $\zeta(x) = (\zeta_m/2)\cos 2\pi x/a_2$ with a maximum corrugation amplitude ζ_m of 0.05 Å. Note that W(112) for which the close-packed row spacing is 4.47 Å as compared to 3.52 Å for Ni(110) shows a considerably larger corrugation. This illustrates the inability of the metal electrons to smooth out the surface as the lattice constant increases.

This is illustrated even more clearly for the case of Au(110) for which the clean surface shows a (1×2) reconstruction /9.11/. Although no definitive structural model for this surface has yet been presented, the missing-row model /9.11/ has been established for Ir(110) /9.12/ which shows the same reconstruction. If this result also holds for Au(110), the spacing of the close-packed rows in the topmost layer would be 8.14 Å, and the corrugation amplitude of a hard-sphere model of this surface

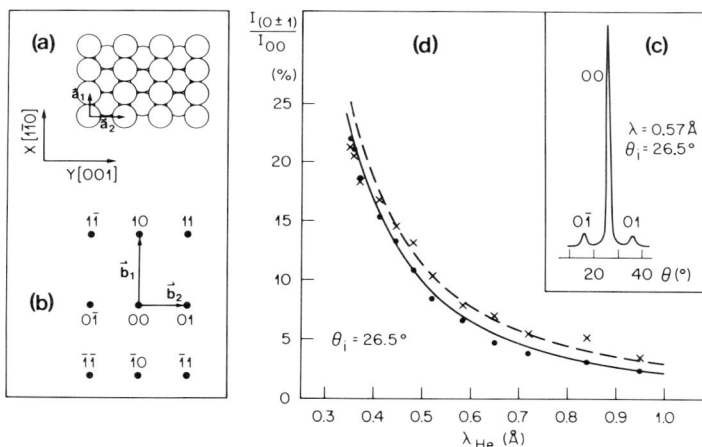

Fig.9.3. (a) Hard-sphere model of the Ni(110) surface showing the unit cell vectors. (b) Reciprocal lattice corresponding to (a). (c) He-diffraction trace for clean Ni(110). (d) He-diffraction intensities of the (01) and (0$\bar{1}$) beams relative to the (00) beam as a function of the wavelength. (01) beam: (•••) experiment; (——) theory. (0$\bar{1}$) beam: (xxx) experiment; (---) theory /9.8/

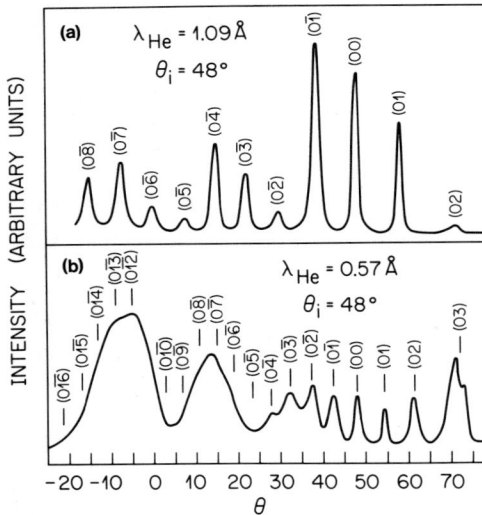

Fig.9.4a,b. He-diffraction traces for Au(110) (1×2) at a surface temperature of 100 K with θ_i = 48°. The wavelength λ_i is (a) 1.09 Å and (b) 0.57 /2.13/

would have a maximum amplitude of 2.3 Å. Results obtained in the authors' laboratory for the diffraction from this surface for λ_i = 0.57 Å and λ_i = 1.09 Å are shown in Fig.9.4 /2.13/. Although we have not been able to extract a corrugation function from these diffraction traces due to strong resonant scattering effects, the corrugation amplitude can be estimated from the rainbow scattering observed for λ = 0.57 Å to be approximately 1.5 Å. This is a substantial fraction of the corrugation amplitude expected on the basis of a hard-sphere model. In the sequence Ni(110), W(112), and Au(110) (1×2), all of which have a similar structure and exhibit a one-dimensional corrugation, the roughness increases considerably, in particular upon going to the reconstructed Au(110) surface. This is accompanied by an inability of the metal electrons to smooth out the corrugation; whereas for the Ni(110) surface the corrugation is about 10% of that expected for the hard-sphere model it is more than 60% of the hard-sphere value on the (1×2) reconstructed Au(110) surface.

9.4   Helium Diffraction from Stepped Metal Surfaces

High-index planes of metal surfaces generally tend to form stepped configurations in which the terraces of low-index orientation are separated by steps of monoatomic height /9.13/. As was discussed above, the terrace will appear very smooth when viewed with atom diffraction, so that diffractive scattering, if observed, will give information about the electron-charge distribution at the step. This is information not previously available with other techniques such as LEED and which could give some insight into the enhanced catalytic activity on stepped surfaces as has been seen, for example, in the H_2-D_2 exchange reaction on Pt(997) /9.2/.

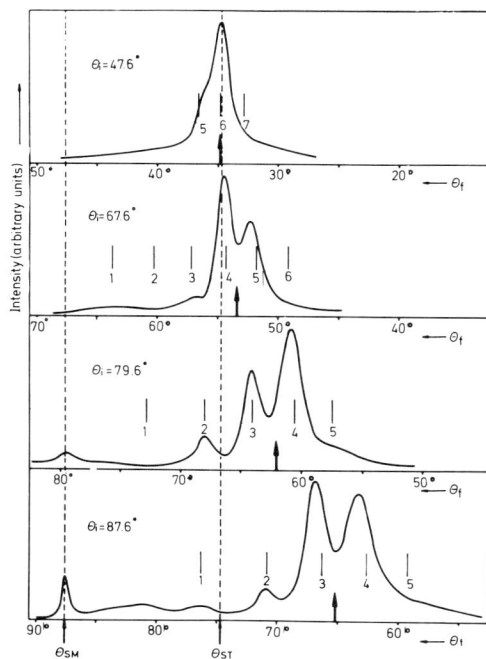

Fig.9.5. He-diffraction traces for different angles of incidence from Pt(997) with the beam incident perpendicular to the step edge and in the step-down direction /9.16/

Helium scattering from stepped surfaces was first reported by CEYER et al. /9.14/. Due to the effusive source used and the limited angular resolution of the detector, these authors were able to observe only the envelope of the diffraction peaks rather than the individual peaks. LAPUJOULADE and LEJAY /9.15/ first published well-resolved diffraction traces for He diffraction from Cu(117), and COMSA et al. /9.16/ have reported investigations of He diffraction from Pt(997). Figure 9.5 shows He-diffraction traces from Pt(997) which consists of (111) terraces with a width distribution centred around 20 Å which are separated by steps of 2.27 Å height. This height difference corresponds to the (111) interplanar spacing. Generally, several diffraction peaks distributed about the specular direction with respect to the terrace are observed. However, for angles very near grazing, a peak corresponding to the specular direction of the macroscopic crystal is also observed as is seen for θ_i = 87.6°. The fact that a beam corresponding to the macroscopic specular direction is seen, suggests that there is a rounding of the electron distribution at the step edge /9.16/ which gives some insight into the electron distribution for these surfaces.

An intensity analysis using a hard-corrugated-wall model has been carried out for the diffraction from Cu(117) /2.12/ for which the mean terrace width is 9 Å and the interplanar spacing of the (100) terraces is 1.81 Å. A preliminary corrugation surface which best fits the experimental intensities is shown in Fig.9.6 /2.12/. It is seen that the abrupt termination of the terrace is smoothed out by the electron-charge distribution. However, the agreement between the experimentally measured and

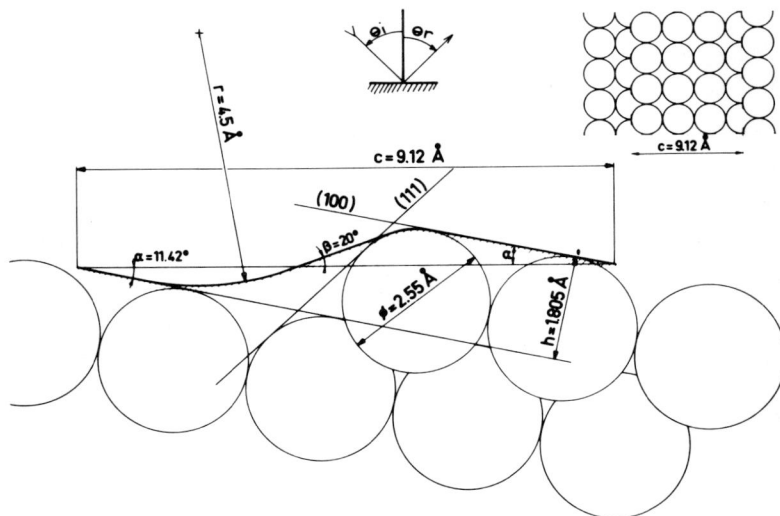

Fig.9.6. Hard-sphere model of the Cu(117) surface together with the best-fit corrugation function obtained from He diffraction spectra /9.16/

the calculated results is not sufficiently good for the model of Fig.9.6 to be accepted as final. The reason for the difficulty in obtaining agreement between experiment and theory on the stepped surfaces is at present not clear. Experimentally, a peak broadening relative to the direct beam is observed for He diffraction from both Cu(117) and Pt(997), which indicates that a certain amount of disorder such as kinks is present on these surfaces. Similarly, there may be a distribution in terrace widths about the mean values cited which will also broaden the peaks and influence the relative intensities of the diffracted beams observed. Both of these effects are complicating features which at present appear to limit our detailed understanding of the electron distribution on stepped surfaces.

10. Structural Studies on Adsorbate-Covered Surfaces

10.1 Introduction

The structural investigation of adsorbate-covered surfaces is the most recent application of atom-diffraction techniques. Information can be gained about both the surface binding site and the electron distribution in the chemisorption bond. Whereas information on the binding site can also be obtained with LEED and ion-scattering experiments, at present no other techniques can probe the electron distribution at surfaces. Metal surfaces, which are the most important substrates for adsorption studies, are particularly well suited for such investigations, since in most cases the corrugation of the clean surface is extremely small (see the previous section). Adsorption can lead to a large change in the corrugation so that the binding site and the electron distribution at the site are clearly seen in favourable cases. Initial experiments were carried out for CH_3OH and H_2O adsorption on NaF /10.1/ and for H_2 adsorption on W(100) /10.2/. In both these cases, diffraction features were observed, but no structural models consistent with the results were proposed. More recently, the authors have carried out extensive investigations of the dissociative adsorption of H_2 /9.8;10.3,4/ and O_2 /10.5/ on Ni(110). Oxygen adsorption on Cu(110) has been studied by LAPUJOULADE et al. /10.6/. In the following, we shall discuss these results and the structural information gained from them.

10.2 Hydrogen Adsorption on Ni(110)

Hydrogen adsorption is important because of its model character as the simplest adsorbate atom, its importance in catalysis and energy storage, and because little is known about the structure of hydrogen adlayers. To date, only the LEED analysis has been published for an adsorbed hydrogen phase /10.7/ primarily because of the low electron scattering power of hydrogen atoms. Since atom scattering will be equally sensitive to all elements, it is ideally suited to the study of gas adsorption and in particular to hydrogen overlayers.

We have found that the dissociative adsorption of H_2 on Ni(110) at 100 K leads to a series of ordered phases as the coverage is increased up to saturation corresponding to $\theta_H \sim 1.6$ monolayers. Experimental details are found in /10.3,4/. Figure 10.1 shows He-diffraction traces for in-plane scattering with the beam incident in

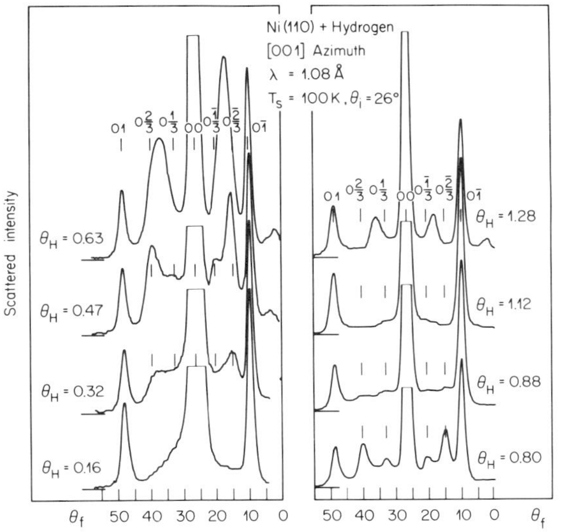

Fig.10.1.
Helium-diffraction traces for various coverages of hydrogen. $T_S = 100$ K, $\theta_i = 26°$, $\lambda_i = 1.08$ Å. The beam is incident in the [001] azimuth. The gain is a factor of two greater for the traces shown on the left-hand side of the figure /10.4/

the [001] azimuth for various hydrogen coverages. For coverages below θ_H = 0.5, (0 ±1/3) and (0 ±2/3) beams are observed. The broad additional peaks seen at θ_H = 0.63 are due to a superposition of intensities from these peaks with that of the (0 ±1/2) beams. At higher coverages, one again observes only (0 ±1/3) and (0 ±2/3) beams which can best be seen near θ_H = 0.8. A further increase in the coverage to θ_H = 1 leads to in-plane scattering with measurable intensity only in the (00) and (0 ±1) beams. Finally, saturation leads to additional intensity in the (0 ±1/2) and (0 ±3/2) beams.

The periodicities of these phases can only be determined by several independent scans through the reciprocal lattice, and such measurements, incorporating in-plane and out-of-plane diffraction /10.4/, show that below θ_H = 0.6 a (2×3) phase is formed. Near θ_H = 0.7, a (2×6) phase is formed which appears to differ in structure from a second (2×6) phase formed at θ_H = 0.8. A further increase in coverage leads to a (2×1) phase at θ_H = 1.0 and a (1×2) phase seen between θ_H = 1.2 and 1.6. Although this adsorption system has been previously investigated with LEED, only the (2×1) and (1×2) surfaces were observed /10.8-11/. Figures 10.2 and 10.3 show in-plane and out-of-plane diffraction traces, respectively, for coverages between θ_H = 0.72 and

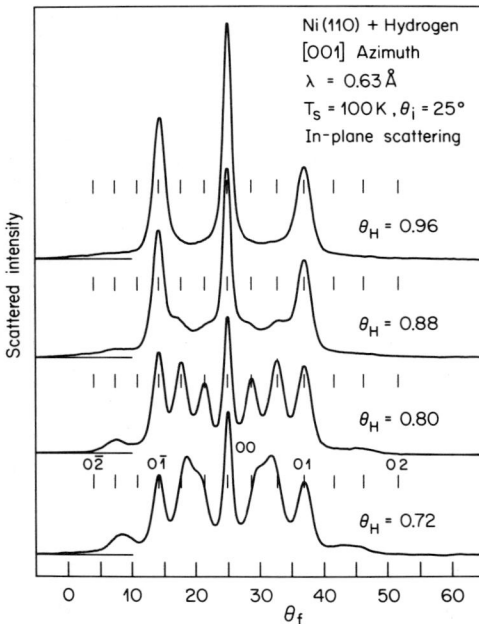

Fig.10.2. In-plane scattered intensity as a function of the scattering angle for hydrogen coverages θ_H between 0.7 and 1. The beam is incident in the [001]azimuth. λ_i = 0.63 Å and θ_i = 25° /10.4/

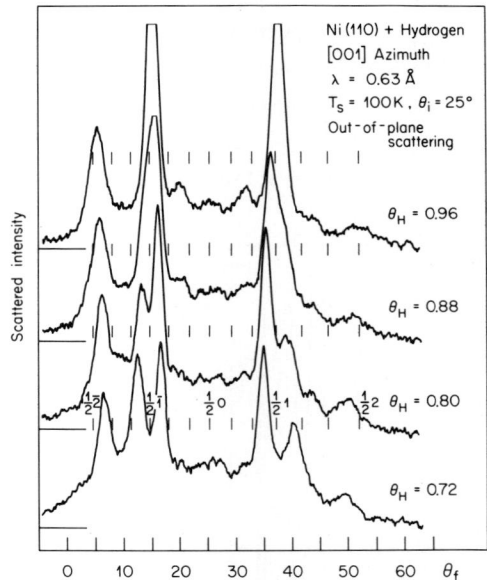

Fig.10.3. Out-of-plane scattered intensity as a function of the scattering angle for θ_H between 0.7 and 1. The vertical scale is magnified by a factor of five relative to that of Fig.10.2. The beam is incident in the [001] azimuth. λ_i = 0.63 Å and θ_i = 25° /10.4/

θ_H = 0.96. The two traces at a given coverage are sufficient to determine the periodicity of the overlayer. These traces taken at small coverage intervals show the
sensitivity of atom diffraction to structural changes on the surface.

We have carried out an analysis of the diffraction intensities for the (2×6) phase
at θ_H = 0.8, the (2×1) phase at θ_H = 1.0, and the (1×2) phase and have been able to
propose structural models consistent with the intensities measured. The high specular intensity observed for θ_H below 0.7 leads us to believe that in this coverage
range, diffraction takes place both from clean and adsorbate-covered patches of the
surface, making an intensity analysis difficult. In analyzing the data, we have not
included a Debye-Waller correction since due to an irreversible surface reconstruction above 220 K, the intensities could not be measured over a wide temperature
range. However, as the substrate temperature was low (100 K) and only small $\underline{G}$ vectors were involved in the diffraction, we believe that corrections for thermal vibrations will not have a significant effect on the results presented below.

In-plane and out-of-plane diffraction traces for the (2×6) phase at θ_H = 0.8 are
shown in Fig.10.4. Note the pronounced rainbow scattering in-plane and the absence
of odd sixth-order peaks in-plane, and of even sixth-order peaks out-of-plane. Also shown in Fig.10.4 is a trace calculated using the corrugation function

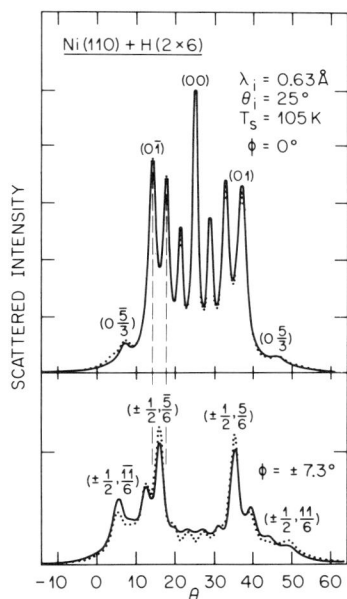

Fig.10.4. Scattered-He intensity as a function of the scattering angle θ for in-
plane and out-of-plane ($\phi \neq 0$) detection for the (2×6) structure. The vertical scale
corresponds to the same intensity for all curves. The (00) beams for the experimental (——) and calculated intensity curve (....) have been set equal. The total elastically scattered intensity is approximately 30% of the incoming intensity /10.3/

$$\zeta(x,y) = -\frac{1}{2}\,\zeta(0\,\tfrac{1}{3})\,\cos2\pi\,\frac{y}{3a_2} - \frac{1}{2}\,\zeta(0\,\tfrac{2}{3})\,\cos2\pi\,\frac{2y}{3a_2} - \frac{1}{2}\,\zeta(01)\,\cos2\pi\,\frac{y}{a_2}$$

$$+\,\zeta(\tfrac{1}{2}\,\tfrac{5}{6})\,\sin2\pi\,\frac{x}{2a_1}\,\sin2\pi\,\frac{5y}{6a_2} \tag{10.1}$$

$$+\,\zeta(\tfrac{1}{2}\,\tfrac{3}{6})\,\sin2\pi\,\frac{x}{2a_1}\,\sin2\pi\,\frac{3y}{6a_2}\quad,$$

where a_1 = 2.49 Å and a_2 = 3.52 Å are the unit-cell lengths of the clean Ni(110) surface in the [001] and [1$\bar{1}$0] azimuths, respectively. To allow a comparison of the experimental with the theoretical results, the latter have been convoluted with the velocity dispersion of the incoming beam and the geometrical resolution of the apparatus. The best-fit parameters are: $\zeta(0\;1/3)$ = -0.09 ± 0.02 Å, $\zeta(0\;2/3)$ = 0.12 ± 0.02 Å, $\zeta(01)$ = 0.09 ± 0.02 Å, $\zeta(1/2\;5/6)$ = 0.06 ± 0.01 Å, and $\zeta(1/2\;3/6)$ = -0.03 ± 0.01 Å. Equation (10.1) contains all coefficients significantly different from zero. These parameters result from a fit to data at three different values of θ_i between 25° and 50° as well as those of Fig.10.4.

The corrugation function given by (10.1) shown in Fig.10.5 has considerably more structure than that for the clean surface shown in Fig.10.6 for comparison, and the maximum corrugation amplitude has increased from 0.05 Å to 0.25 Å. The structure of the corrugation surface can be described by paired zig-zag chains parallel to the [1$\bar{1}$0] direction which are separated by rows of isolated maxima.

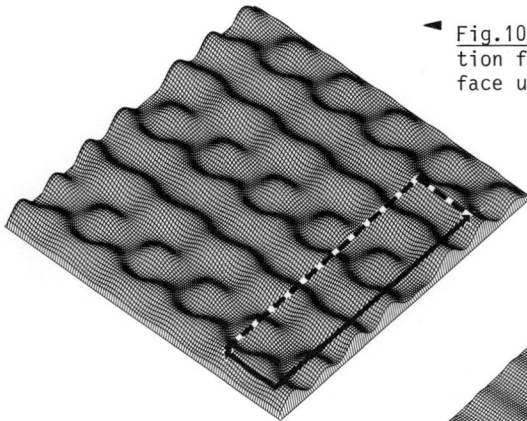

Fig.10.5. Best-fit corrugation function for the (2×6) phase. The surface unit cell is indicated /10.3/

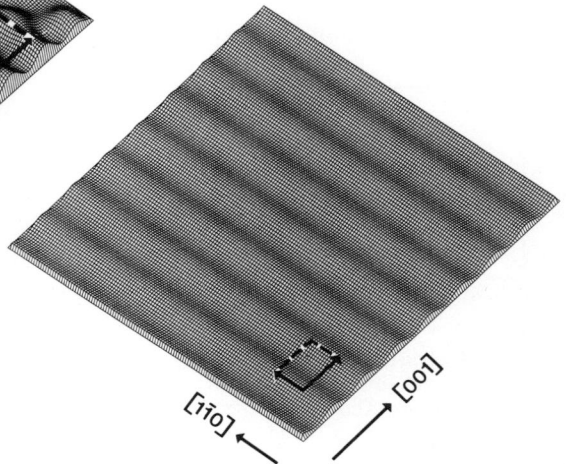

Fig.10.6. Best-fit corrugation function for the clean Ni(110) surface. The surface unit cell is indicated

Since the corrugation function shown in Fig.10.5 is a replica of the electron-charge density (Chap.2), the maxima can be attributed to the localization of charge induced by the chemisorption bond and to the position of the underlying hydrogen atom. To deduce the positions of the hydrogen atoms on the surface, the registry of the corrugation function to the nickel lattice must be known. The registry along the [001] azimuth is fixed through the still visible corrugation due to the substrate shown by the magnitude of $\zeta(01)$, but that along the [1$\bar{1}$0] azimuth cannot be uniquely determined due to the weak substrate corrugation along this direction, and we have chosen the configuration yielding equivalent adsorption sites of the highest coordination. With this choice, we arrive at the structural model shown in Fig.10.7.

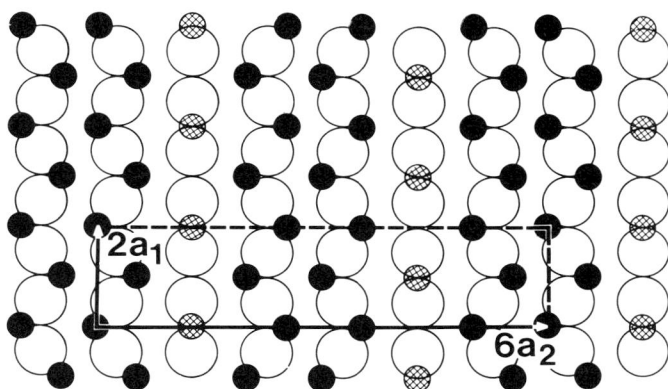

Fig.10.7. Hard-sphere model for the (2×6) adsorption phase. The small filled and shaded circles represent H-atoms, and the large circles, the outermost layer of the Ni(110) substrate. The surface unit cell is indicated /10.3/

It corresponds to a distorted hexagonal close packing of the hydrogen atoms. Note that three- and two-fold sites are occupied and that the height of the corrugation function differs for the two sites. The lower height of 0.18 Å for the two-fold site compared with 0.25 Å for the three-fold site is consistent with the coordination dependence of bond lengths /10.11/.

An increase in the coverage to θ_H = 1 leads to the in-plane and out-of-plane diffraction traces shown in Fig.10.8. The absence of the ($\pm$1/2 0) beams for all wavelengths and at all angles of incidence investigated shows that the unit cell is non-primitive and contains glide lines parallel to the [1$\bar{1}$0] azimuth. The best-fit corrugation function, consistent with the glide lines and the intensities of diffraction traces at seven angles of incidence between 25° and 50° with the beam incident in both the [001] and [1$\bar{1}$0] azimuths, is given by

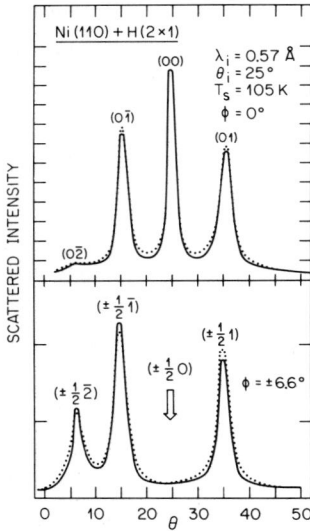

Fig.10.8. Scattered-He intensity as a function of the scattering angle θ_f for in-plane ($\phi = 0°$) and out-of-plane ($\phi \neq 0°$) detection for the (2x1) structure. The vertical scale corresponds to the same intensity for all curves. The (00) beams for the experimental (——) and calculated intensity curves (....) have been set equal. The total elastically scattered intensity is approximately 30% of the incoming intensity /10.3/

$$\zeta(x,y) = -\frac{1}{2}\,\zeta(01)\,\cos2\pi\,\frac{y}{a_2} - \frac{1}{2}\,\zeta(02)\,\cos2\pi\,\frac{2y}{a_2}$$
$$+ \zeta(\tfrac{1}{2}\,1)\,\sin2\pi\,\frac{x}{2a_1}\,\sin2\pi\,\frac{y}{a_2} \qquad\qquad (10.2)$$
$$+ \zeta(\tfrac{1}{2}\,2)\,\sin2\pi\,\frac{x}{2a_1}\,\sin2\pi\,\frac{2y}{a_2}\quad.$$

The best-fit parameters are: $\zeta(01) = 0.10 \pm 0.02$ Å, $\zeta(02) = 0.03 \pm 0.01$ Å, $\zeta(1/2\ 1) = 0.07 \pm 0.01$ Å, and $\zeta(1/2\ 2) = 0.02 \pm 0.01$ Å, and the maximum corrugation amplitude is 0.26 ± 0.02 Å. Equation (10.2) contains all coefficients significantly different from zero. The dotted trace in Fig.10.8 is calculated using (10.2) in the same way as was described for the (2×6) phase.

The best-fit corrugation function given by (10.2) is shown in Fig.10.9. The zig-zag chains present in the (2×6) phase now cover the entire surface, and determining the registry as before, we arrive at the structural model shown in Fig.10.10. The structure again corresponds to a distorted hexagonal close packing of the hydrogen atoms and maximizes the distance between the adatoms.

Note that for both the (2×6) and (1×2) phases analysed, the coverage can be directly determined by counting the number of adatoms per unit cell. In the (2×6) phase, there are ten adatoms per 12 Ni atoms in the topmost substrate layer so that $\theta_H = 0.83$. In the (2×1) phase, there are two adatoms in the unit cell which comprises two Ni atoms; therefore the coverage is $\theta_H = 1$. Relying on $\theta_H = 1$ for the (2×1) phase, the absolute coverages of the other phases could be determined by flash desorption

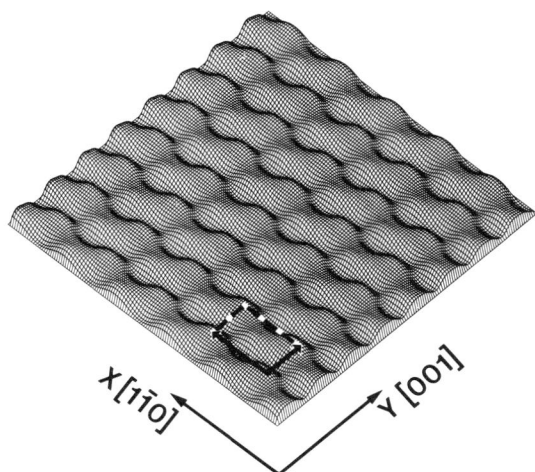

Fig.10.9. Best-fit corrugation function for the (2×1) phase. The surface
unit cell is indicated /10.3/

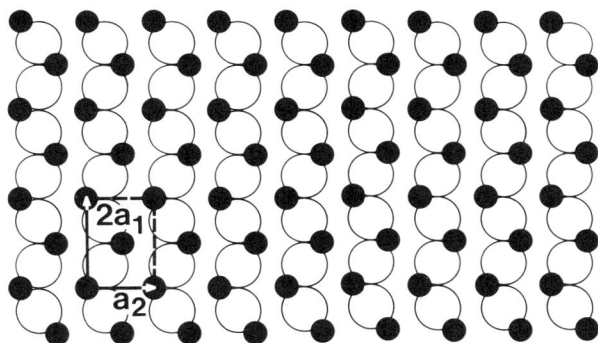

Fig.10.10. Hard-sphere model for the (2×1) adsorption phase. The small filled circles
represent H-atoms, and the larger circles, the outermost layer of the Ni(110) substrate. The surface unit cell is indicated /10.3/

measurements. These values are cited in the text and in the figures. The absolute
coverage obtained in this way for the (2×6) structure discussed above is θ_H =
0.8 ± 0.05. This is in very good agreement with the result of the structural analysis θ_H = 0.83 and proves the internal consistency of the coverage determination.

Both the (2×6) and (2×1) phases show localized maxima in the corrugation function which correspond to the position of the underlying hydrogen atom. As seen from
Fig.10.10, in the (2×1) phase all equivalent sites are filled and an increase in
the coverage from θ_H = 1 to 1.6 must lead to an extensive rearrangement of the surface. On the basis of the intensity of the LEED beams, it can be concluded that the
surface reconstructs in this coverage range, and a dynamical LEED calculation came
to the tentative conclusion that a pairing of the nickel rows occurs /10.9/.

Our helium-diffraction results for the (1×2) phase show that the corrugation function is very weakly structured in the $[1\bar{1}0]$ azimuth. Along the $[001]$ direction, $\zeta(\underline{R})$ can be written as

$$\zeta(x) = \frac{\zeta_1}{2} \cos 2\pi \frac{x}{a_2} + \frac{\zeta_2}{2} \cos 4\pi \frac{x}{a_2} + \frac{\zeta_3}{2} \cos 6\pi \frac{x}{a_2} + \frac{\zeta_5}{2} \cos 10\pi \frac{x}{a_2} \quad . \tag{10.3}$$

The best-fit parameters are $\zeta_1 = 0.20 \pm 0.02$ Å, $\zeta_2 = 0.18 \pm 0.02$ Å, $\zeta_3 = 0.04 \pm 0.01$ Å, and $\zeta_5 = -0.002 \pm 0.001$ Å, and the maximum corrugation amplitude is 0.325 ± 0.03 Å. The corrugation function corresponding to (10.3) is shown in Fig.10.11a. It is struc-

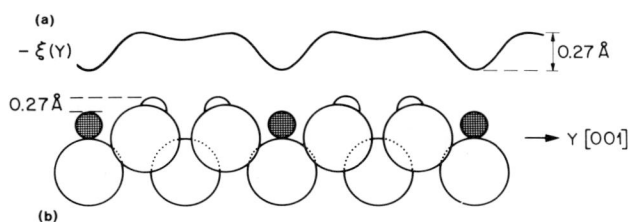

Fig.10.11. (a) Corrugation function for the (1×2) phase together with (b) the structural model proposed. The small open and shaded circles represent hydrogen atoms, and the open circles the outermost layer of the Ni(110) substrate /10.4/

tured only in the $[001]$ azimuth and no evidence for localized adsorption sites is seen which is consistent with the surface reconstruction proposed. The lack of localized maxima makes a structural assignment less clearcut that for the (2×1) and (2×6) phases. However, a model consistent with the experimentally determined coverage, the (1×2) periodicity observed, and the corrugation function of Fig.10.11a is shown in Fig.10.11b. This model incorporates the nickel-row pairing proposed by DEMUTH /10.9/ which allows hydrogen to be adsorbed in the second layer. The small corrugation in the $[1\bar{1}0]$ azimuth suggests that either the electron distribution about a hydrogen atom is highly anisotropic, or that the partially delocalized metal electrons rather than the hydrogen electrons extend further from the surface into the vacuum. The striking difference between the localized maxima in the corrugation function found for the (2×1) and (2×6) phases and the one-dimensional form of $\zeta(\underline{R})$ for the (1×2) phase shows the type of information which can be obtained about the structural dependence of the chemisorption bond.

10.3 Oxygen Adsorption on Ni(110)

Oxygen adsorption on Ni(110) has been studied with a number of techniques in recent years /10.10-15/. Of particular interest is whether oxygen is adsorbed on top of the substrate, in which case a two-dimensional corrugation with localized maxima should be observed with atom diffraction, or whether oxygen induces a surface reconstruction. If the reconstruction is such that the oxygen is screened by the metal electrons [as in the (1×2) phase of H_2 on Ni(110)], no localized maxima will be seen in $\zeta(\underline{R})$.

We have observed He diffraction from three ordered phases of O_2 on Ni(110) which have (3×1), (2×1), and again (3×1) periodicities as the coverage increases /10.5/. These phases were studied previously in our laboratory using LEED, AES, and SIMS /10.14/. All of these phases are strongly corrugated in the $[1\bar{1}0]$ azimuths and very weakly corrugated in the [001] azimuths. The absence of localized maxima strongly suggests that the oxygen is incorporated into the surface.

Diffraction traces for the (2×1) phase are shown in Fig.10.12 for two angles of incidence. Strong oscillations of the diffraction intensities with θ_i are observed as seen for the (00) and $(\bar{2}0)$ beams in Figs.10.13,14, respectively. Note that minima in the (00) intensity appear as maxima in the $(\bar{2}0)$ intensity and vice versa as indicated by the arrows in Fig.10.14. (See Sect.5.4 for a discussion of this behaviour.)

These oscillations in intensity make calculation of $\zeta(\underline{R})$ difficult, but valuable structural insight can be gained alone by the observation that the corrugation function for this phase is strongly corrugated in the $[1\bar{1}0]$ azimuth and weakly corrugated in the [001] azimuth. Figure 10.15 shows three models which have been suggested by other investigators for this phase. That shown in Fig.10.15a based on a LEED calculation of DEMUTH /10.9/ corresponds to on-top chemisorption and is hardly consistent with a one-dimensional corrugation. The model of GERMER and MACRAE /10.16/ shown in Fig.10.15b includes a surface reconstruction, but would probably also correspond to a two-dimensional corrugation. The model of VAN DEN BERG et al. /10.15/ shown in Fig.10.15c based on ion-scattering experiments appears most consistent with our results, since the oxygen atoms will be screened at the sites indicated, and the corrugation in the [001] azimuth should be very small. Hard-wall calculations for small angles of incidence, where the resonant scattering is not as pronounced, are in progress /10.5/ and should yield more detailed structural information.

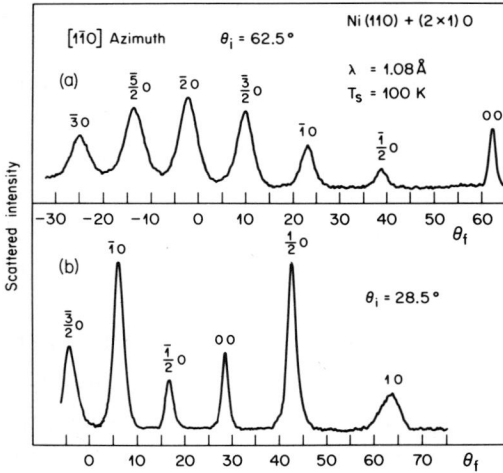

Fig.10.12a,b. In-plane scattered intensity as a function of the scattering angle θ_f for the (2×1) phase of oxygen on Ni(110) for (a) $\theta_i = 62.5°$ and (b) $\theta_i = 28.5°$. The beam is incident in the [1$\bar{1}$0] azimuth. $T_s = 100$ K. $\lambda_i = 1.08$ Å

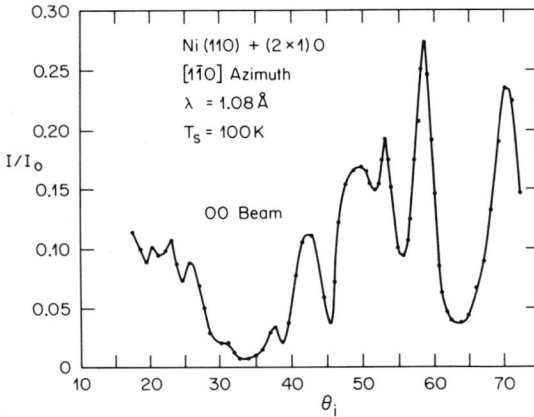

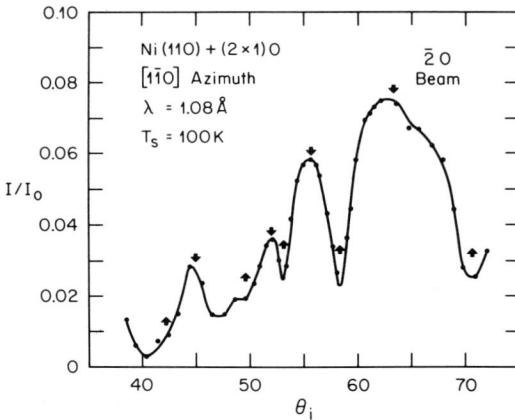

Fig.10.13. Specular-beam intensity as a function of θ_i for the (2×1) phase of oxygen on Ni(110). The beam is incident in the [1$\bar{1}$0] azimuth. $T_s = 100$ K, $\lambda_i = 1.08$ Å /10.5/

Fig.10.14. Intensity of the ($\bar{2}$0) beam as a function of θ_i for the (2×1) phase of oxygen on Ni(110). The geometry and wavelength are as for Fig.10.13. The upward- and downward-pointing arrows show the angle of incidence at which the (00) beam intensity has maxima and minima, respectively. (Fig.10.13) /10.5/

Ni(110) + (2×1)O

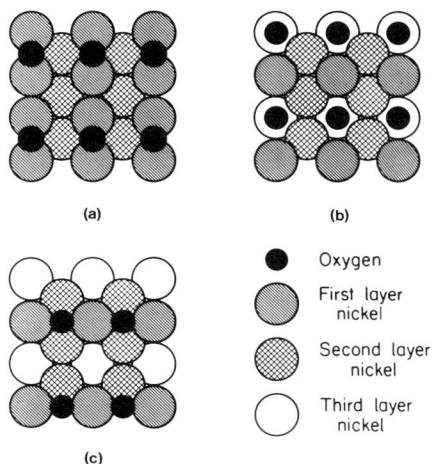

(a)   (b)

(c)

● Oxygen

◯ First layer
  nickel

▦ Second layer
  nickel

◯ Third layer
  nickel

Fig.10.15a-c. Structural models proposed for
the (2×1) phase of oxygen on Ni(110). a) On-
top chemisorption /10.9/, (b) reconstruction
involving replacement of nickel atoms by
oxygen /10.16/, (c) reconstruction involving
a movement of nickel atoms /10.15/

10.4 Oxygen Adsorption on Cu(110)

Helium diffraction has also been observed for the (2×1) and (6×2) phases of oxygen
on Cu(110) /10.6/. As for oxygen on Ni(110), many diffraction peaks are observed in-
dicating a large corrugation. The experiments were carried out only with the beam
incident in the $[1\bar{1}0]$ azimuth and only with in-plane detection, so that no direct
information about the corrugation in the $[001]$ azimuth is available. Indirect evidence that the latter corrugation is weak is provided by the observation that the
resonant scattering observed can be assigned to in-plane $\underline{G}$ vectors only. However,
more experiments are necessary to verify this preliminary conclusion. LAPUJOULADE
et al. /10.6/ have fit the data to a one-dimensional corrugation function shown in
Fig.10.16. The radii R and r were set equal to the Van der Waal's radii of the oxygen and helium atoms, respectively, and d was treated as an adjustable parameter.
Qualitative agreement between the theoretically calculated and experimentally de-

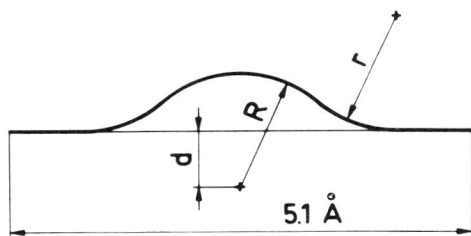

5.1 Å

Fig.10.16. Corrugation function proposed for the (2×1) phase of oxygen
on Cu(110). R and r are the Van der
Waal's radii of O and He, respectively. The best-fit parameter $d = 0.7 \pm 0.1$ Å
/10.6/

termined intensities was obtained for d = 0.7 ± 0.1 Å. However, the agreement is not nearly as good as that obtained for H_2 adsorption on Ni(110) (see Sect.10.2), so that this model must be regarded as preliminary.

Acknowledgements. We wish to thank W. Brenig, M. Cardillo, and W. Schlup for generously providing material to us prior to its publication. We have learned a lot from discussions with A. Baratoff, V. Celli, and N. Garcia. Our own investigations have benefitted greatly through the skillful experimental assistance of W. Stocker.

We should also like to thank the Publications Department of the IBM Zurich Research Laboratory, specifically U. Bitterli, Dilys Brüllmann, Dianne Kunz (especially), and M. Wagner.

References

1.1   F. Jona: J. Phys. C *11*, 4271 (1978)
1.2   J. Estermann, O. Stern: Z. Phys. *61*, 95 (1930)
1.3   G. Brusdeylins, R.B. Doak, J.P. Toennies: Phys. Rev. Lett. *44*, 1417 (1980)
1.4   J.E. Hurst, C.A. Becker, J.P. Cowin, K.C. Janda, L. Wharton, D.J. Auerbach: Phys. Rev. Lett. *43*, 1175 (1979)
1.5   T. Engel: J. Chem. Phys. *69*, 373 (1978)
1.6   T. Engel, G. Ertl: J. Chem. Phys. *69*, 1267 (1978)
1.7   J.P. Toennies: Appl. Phys. *3*, 91 (1974)
1.8   M.W. Cole, D.R. Frankl: Surf. Sci. *70*, 585 (1978)
1.9   H. Wilsch: *Topics in Surface Chemistry*, ed. by E. Kay, P. Bagus (Plenum Press, New York 1978) p. 135
1.10 H. Hoinkes: Rev. Mod. Phys. *52*, 933 (1980)
1.11 K.H. Rieder, T. Engel: Nucl. Instrum. Methods *170*, 483 (1980)
1.12 T. Engel, K.H. Rieder: *Proc. 4th Int. Conf. on Solid Surfaces and 3rd European Conf. on Surface Science*, Cannes, 1980, Suppl. à la Revue "Le Vide, les Couches Minces" Nr. 201, p. 801 (1980)
1.13 G. Armand, J. Lapujoulade: *Proc. 11th Rarefied Gas Dynamics Symposium*, Cannes, 1978, ed. by R. Campargue (Commisariat à L'Energie Atomique, Paris 1979) p. 1329
1.14 F.O. Goodman: CRC Crit. Rev. in Sol. State and Mat. Sci. *7*, 33 (1977)
1.15 F.O. Goodman, H.Y. Wachmann: *Dynamics of Gas-Surface Scattering* (Academic Press, New York 1976)

2.1   R.E. Stickney: *The Structure and Chemistry of Solid Surfaces*, ed. by G.A. Somorjai (John Wiley, New York 1969)
2.2   H.U. Finzel, H. Frank, H. Hoinkes, M. Luschka, H. Nahr, H. Wilsch, U. Wonka: Surf. Sci. *49*, 577 (1975)
2.3   H. Hoinkes, L. Greiner, H. Wilsch: *Proc. 7th Int. Vacuum Congr. and 3rd Int. Conf. Solid Surfaces*, Vienna, 1977, ed. by R. Dobrozemsky et al. (American Institute of Aeronautics and Astronautics, New York 1977) p. 1349
2.4   H. Frank, H. Hoinkes, H. Wilsch: Surf. Sci. *63*, 121 (1977)
2.5   E. Ghio, L. Mattera, C. Salvo, F. Tommasini, U. Valbusa: J. Chem. Phys. *73*, 556 (1980)
2.6   G. Derry, D. Wesner, S.V. Krishnaswamy, D.R. Frankl: Surf. Sci. *74*, 245 (1978)

2.7 G. Derry, D. Wesner, S.V. Krishnaswamy, M.W. Cole, D.R. Frankl: *Proc. 7th Int. Vacuum Congr. and 3rd Int. Conf. Solid Surfaces*, Vienna, 1977, ed. by R. Dobrozemsky et al. (American Institute of Aeronautics and Astronautics, New York 1977) p. 1353

2.8 G. Derry, D. Wesner, W. Carlos, D.R. Frankl: Surf. Sci. *87*, 629 (1979)

2.9 G. Boato, P. Cantini, C. Guidi, R. Tatarek, G.P. Felcher: Phys. Rev. B *20*, 3957 (1979)

2.10 P. Cantini, G.P. Felcher, R. Tatarek: *Proc. 7th Int. Vacuum Congr. and 3rd Int. Conf. Solid Surfaces*, Vienna, 1977, ed. by R. Dobrozemsky et al. (American Institute of Aeronautics and Astronautics, New York 1977) p. 1357

2.11 P. Cantini, R. Tatarek, G.P. Felcher: Phys. Rev. B *19*, 1161 (1979)

2.12 J. Lapujoulade, Y. Lejay, N. Papanicolaou: Surf. Sci. *90*, 133 (1979)

2.13 K.H. Rieder, T. Engel, N. Garcia: *Proc. 4th Int. Conf. on Solid Surfaces and 3rd European Conf. on Surface Science*, Cannes, France, Sept. 1980, Suppl. à la Revue "Le Vide, les Couches Minces" Nr. 201, p. 861

2.14 J.M. Soler, V. Celli, N. Garcia, K.H. Rieder, T. Engel: Surf. Sci. *108*, 1 (1981)

2.15 R.J. Le Roy: Surf. Sci. *59*, 541 (1976)

2.16 L. Mattera, F. Rosatelli, C. Salvo, F. Tommasini, U. Valbusa, G. Vidali: Surf. Sci. *93*, 515 (1980)

2.17 J.A. Barker: *Rare Gas Solids*, Vol. 1, ed. by M.L. Klein, J.A. Venables (Academic Press, New York 1976)

2.18 W.A. Steele: *The Interaction of Gases with Solid Surfaces* (Pergamon Press, Oxford 1974)

2.19 A. Tsuchida: Surf. Sci. *14*, 375 (1969)

2.20 A. Tsuchida: Surf. Sci. *46*, 611 (1974)

2.21 J.M. Rogowska: J. Chem. Phys. *68*, 3910 (1978)

2.22 P.G. Hall, M.A. Rose: Surf. Sci. *74*, 644 (1978)

2.23 E.M. Lifshitz: Sov. Phys.-JETP *2*, 73 (1956). (For $z > 100$ Å, radiation effects change the asymptotic behaviour to $V_{as} \sim z^{-4}$.)

2.24 E. Zaremba, W. Kohn: Phys. Rev. B *13*, 2270 (1976)

2.25 L.W. Bruch, H. Watanabe: Surf. Sci. *65*, 619 (1977)

2.26 G. Vidali, M.W. Cole, C. Schwartz: Surf. Sci. *87*, L273 (1979)

2.27 U. Landman, G.G. Kleinman: J. Vac. Sci. Technol. *12*, 206 (1975)

2.28 D.L. Freeman: J. Chem. Phys. *62*, 941 (1975)

2.29 E. Zaremba, W. Kohn: Phys. Rev. B *15*, 1769 (1977)

2.30 J.E. van Himbergen, R. Silbey: Solid State Commun. *23*, 623 (1977)

2.31 N.D. Lang, A.R. Williams: Phys. Rev. B *18*, 616 (1978)

2.32 N. Esbjerg, J.K. Norskov: Phys. Rev. Lett. *45*, 807 (1980)

2.33 L. Mattera, C. Salvo, S. Terreni, F. Tommasini: Surf. Sci. *97*, 158 (1980)

2.34 H. Hoinkes, H. Wilsch: *Proc. 4th Int. Conf. on Solid Surfaces and 3rd European Conf. on Surface Science*, Cannes, Sept. 1980, Suppl. à la Revue "Le Vide, les Couches Minces" Nr. 201, p. 87 (1980)

2.35 W.E. Carlos, M.W. Cole: Surf. Sci. *91*, 339 (1980)

2.36 G. Boato, P. Cantini, R. Tatarek: Phys. Rev. Lett. *40*, 887 (1978)

2.37 G. Boato, P. Cantini, R. Tatarek, G.P. Felcher: Surf. Sci. *80*, 518 (1979)

2.38 W.E. Carlos, M.W. Cole: Phys. Rev. B *21*, 3713 (1980)

3.1 J.W. Strutt (Lord Rayleigh): *The Theory of Sound*, Vol. 2 (Macmillan, London, 1896) p. 272

3.2 U. Garibaldi, A.C. Levi, R. Spadacini, G.E. Tommei: Surf. Sci. *48*, 649 (1975)

3.3 R.I. Masel, R.P. Merrill, W.H. Miller: Phys. Rev. B *12*, 5545 (1975)

3.4 R.I. Masel, R.P. Merrill, W.H. Miller: J. Chem. Phys. *65*, 2690 (1976)

3.5 F.O. Goodman: J. Chem. Phys. *66*, 976 (1977)

3.6 F. Toigo, A. Marvin, V. Celli, N.R. Hill: Phys. Rev. B *15*, 5618 (1977)

3.7 N. Garcia, N. Cabrera: *Proc. of the 3rd Int. Conf. on Solid Surfaces*, Vol. 1, Vienna, 1977, ed. by R. Dobrozemsky et al. (Berger, Vienna, 1977) p. 379

3.8 N. Garcia, N. Cabrera: Phys. Rev. B *18*, 576 (1978)

176

3.9   G. Armand, J.R. Manson: Phys. Rev. B *18*, 6510 (1978)
3.10 G. Armand, J. Lapujoulade, J.R. Manson: Surf. Sci. Lett. *82*, 2625 (1979)
3.11 N. Garcia: Surf. Sci. *71*, 220 (1978)
3.12 R. Petit, M. Cadilhac: C.R. Acad. Sci. Ser. B *262*, 468 (1966)
3.13 R.F. Millar: Proc. Cambridge Philos. Soc. *69*, 217 (1971)
3.14 N.R. Hill, V. Celli: Phys. Rev. B *17*, 2478 (1978)
3.15 P.M. Van den Berg, J.T. Fokkema: J. Opt. Soc. Am. *69*, 27 (1979)
3.16 N. Garcia, J. Ibanez, J. Solana, N. Cabrera: Surf. Sci. *60*, 385 (1976)
3.17 N. Garcia, J. Ibanez, J. Solana, N. Cabrera: Solid State Commun. *20*, 1559 (1976)
3.18 N. Garcia: Phys. Rev. Lett. *37*, 912 (1976)
3.19 N. Garcia: J. Chem. Phys. *67*, 897 (1977)
3.20 G. Boato, P. Cantini, L. Mattera: Surf. Sci. *55*, 191 (1976)
3.21 H. Chow, E.D. Thompson: Surf. Sci. *54*, 269 (1976)
3.22 K.H. Rieder, A. Baratoff, U.T. Höchli: Surf. Sci. *100*, L475 (1980)
3.23 C. Lopez, F.J. Yndurain, N. Garcia: Phys. Rev. B *18*, 970 (1978)
3.24 W. Schlup: private communication
3.25 J.D. McClure: J. Chem. Phys. *52*, 2712 (1970)
3.26 T. Engel, K.H. Rieder: Surface Sci., to be published
3.27 H. Lipson, W. Cochran: *The Determination of Crystal Structures* (G. Bell and Sons Ltd., London 1957)
3.28 K.H. Rieder, N. Garcia, V. Celli: Surf. Sci. *108*, 169 (1981)
3.29 G. Armand, J.R. Manson: Phys. Rev. Lett. *43*, 1839 (1979)
3.30 G. Armand, J. Lapujoulade, J. Lejay: *Proc. 4th Int. Conf. on Solid Surfaces and 3rd European Conf. on Surface Science*, Cannes, France, Sept. 1980, Suppl. à la Revue "Le Vide, les Couches Minces" Nr. 201 (1980) p. 857

4.1   P. Debye: Ann. Phys. (Leipzig) *43*, 49 (1914)
4.2   I. Waller: Z. Phys. *17*, 398 (1923)
4.3   R.W. James: *The Optical Principles of the Diffraction of X-Rays* (Bell, London 1967)
4.4   A.C. Levi, H. Suhl: Surf. Sci. *88*, 221 (1979)
4.5   J.L. Beeby: J. Phys. C *4*, L359 (1971)
4.6   H. Hoinkes, H. Nahr, H. Wilsch: Surf. Sci. *33*, 516 (1972); *40*, 457 (1973)
4.7   G. Armand, J. Lapujoulade, Y. Lejay: Surf. Sci. *63*, 143 (1977)
4.8   G. Armand, J.R. Manson: Surf. Sci. *80*, 532 (1979)
4.9   G. Armand: J. Phys. (Paris) *38*, 989 (1977)
4.10 H. Hoinkes, H. Nahr, H. Wilsch: Surf. Sci. *30*, 363 (1972)
4.11 J. Lapujoulade, Y. Lejay, G. Armand: Surf. Sci. *95*, 107 (1980)
4.12 J. Böheim, W. Brenig, J. Stutzki: Z. Phys., to be published
4.13 J. Böheim, W. Brenig: Z. Phys. *41*, 243 (1981)
4.14 F.O. Goodman: *Progress in Surface Science*, Vol. 5, ed. by S.G. Davisson (1975) p. 261
4.15 H. Asada: Surf. Sci. *81*, 386 (1979)
4.16 E. Müller-Hartmann, T.V. Ramakrischnan, G. Toulouse: Solid State Commun. *9*, 99 (1971)
4.17 B.F. Mason, B.R. Williams: Surf. Sci. *75*, L786 (1978)
4.18 P. Cantini, R. Tatarek: Phys. Rev. B *23*, 3030 (1981)

5.1   N. Garcia, F.O. Goodman, V. Celli, N.R. Hill: Phys. Rev. B *19*, 1808 (1979)
5.2   R.O. Frisch, O. Stern: Z. Phys. *84*, 430 (1933)
5.3   R.O. Frisch: Z. Phys. *84*, 443 (1933)
5.4   J.E. Lennard-Jones, A.F. Devonshire: Nature (London) *137*, 1069 (1936); Proc. R. Soc. London, Ser. A *156*, 6 (1935)
5.5   J.A. Meyers, D.R. Frankl: Surf. Sci. *51*, 61 (1975)
5.6   H. Chow, E.D. Thompson: Surf. Sci. *59*, 225 (1976)
5.7   M.P. Liva, G. Derry, D.R. Frankl: Phys. Rev. Lett. *37*, 1413 (1976)
5.8   W.E. Carlos, G. Derry, D.R. Frankl: Phys. Rev. B *19*, 3258 (1979)

5.9   K.L. Wolfe, J.H. Weare: Phys. Rev. Lett. *41*, 1663 (1978)
5.10 K.L. Wolfe, C.H. Harvie, J.H. Weare: Solid State Commun. *27*, 1293 (1978)
5.11 D.R. Frankl, D. Wesner, S.V. Krishnaswamy, G. Derry, T. O'Gorman:
      Phys. Rev. Lett. *41*, 60 (1978)
5.12 N. Cabrera, V. Celli, F.O. Goodman, J.R. Manson: Surf. Sci. *19*, 67 (1970)
5.13 F.O. Goodman, W.K. Tan: J. Chem. Phys. *59*, 1805 (1973)
5.14 G. Wolken Jr.: J. Chem. Phys. *58*, 3047 (1973); Chem. Phys. Lett. *21*, 373 (1973);
      J. Chem. Phys. *60*, 2210 (1974)
5.15 J.H. Weare, E. Thiele: J. Chem. Phys. *48*, 513 (1968); *48*, 2329 (1968)
5.16 C.H. Harvie, J.H. Weare: Phys. Rev. Lett. *40*, 187 (1978)
5.17 Y. Hamauzu: J. Phys. Soc. Jpn. *42*, 961 (1977)
5.18 H. Chow: Surf. Sci. *66*, 221 (1977)
5.19 H. Chow, E.D. Thompson: Surf. Sci. *82*, 1 (1979)
5.20 K.L. Wolfe, D. Malik, J.H. Weare: J. Chem. Phys. *67*, 1031 (1977)
5.21 V. Celli, N. Garcia, J. Hutchison: Surf. Sci. *87*, 112 (1979)
5.22 N. Garcia, V. Celli, F.O. Goodman: Phys. Rev. B *19*, 634 (1979)
5.23 L. Greisner, H. Hoinkes, H. Kaarmann, H. Wilsch, N. Garcia: Surf. Sci. *94*,
      L195 (1980)
5.24 N. Garcia, W.E. Carlos, M.W. Cole, V. Celli: Phys. Rev. B *21*, 1636 (1980)
5.25 J. Hutchison, V. Celli, N.R. Hill, M. Haller: *Proc. of the 12th Rarefied Gas
      Dynamics Conf.*, Charlottesville, June 1980, in press
5.26 J. Hutchison: Phys. Rev. B *22*, 5671 (1980)
5.27 J. Hutchison, V. Celli: Surf. Sci. *93*, 263 (1980)
5.28 J.M. Soler, V. Celli, N. Garcia, K.H. Rieder, T. Engel: *Proc. 4th Int. Conf.
      on Solid Surfaces and 3rd European Conf. on Surface Science*, Cannes, France,
      Sept. 1980, Suppl. à la Revue "Le Vide, les Couches Minces" Nr. 201 (1980)
      p. 815

6.1   P.L. Redhead, J.P. Hobson, E.V. Kornelsen: *The Physical Basis of Ultrahigh
      Vacuum* (Chapman and Hall, London 1968)
6.2   G.L. Weissler, R.W. Carlson (eds.): *Vacuum Physics and Technology*, Methods
      of Experimental Physics, Vol. 14 (Academic Press, New York 1979)
6.3   G. Ertl, J. Küppers: *Low Energy Electrons and Surface Chemistry* (Verlag Chemie,
      Weinheim 1974)
6.4   T.N. Rhodin, G. Ertl (eds.): *The Nature of the Surface Chemical Bond* (North-
      Holland, Amsterdam 1979)
6.5   J.B. Anderson, R.P. Andres, J.B. Fenn: *Supersonic Nozzle Beams*, Advances in
      Chemical Physics, Vol. 5, ed. by J. Ross (Academic Press, New York 1966)
6.6   M.A.D. Fluendy, K.P. Lawley: *Chemical Applications of Molecular Beam Scatter-
      ing* (Chapman and Hall, London 1973)
6.7   J.B. Anderson: Molecular Beams from Nozzle Sources, in Molecular Beams and
      Low Density Gas Dynamics, ed. by P.P. Wegener (Marcel Dekker, New York 1974)
6.8   H. Pauly, J.P. Toennies: *Methods of Experimental Physics*, 7A, ed. by B. Beder-
      son, W. Fite (Academic Press, New York 1968)
6.9   E.H. Kennard: *Kinetic Theory of Gases* (McGraw-Hill, New York 1938)
6.10 H.V. Hostettler, R.B. Bernstein: Rev. Sci. Instrum. *31*, 872 (1960)
6.11 J. Wykes: J. Phys. E. *2*, 899 (1960)
6.12 J.A. Giormaine, T.C. Wang: J. Appl. Phys. *31*, 463 (1960)
6.13 J.C. Johnson, A.T. Stair, J.L. Pritchard: J. Appl. Phys. *37*, 155 (1966)
6.14 T.H. Johnson: Phys. Rev. *31*, 103 (1928)
6.15 A. Kantrowitz, J. Grey: Rev. Sci. Instrum. *22*, 328 (1951)
6.16 G.B. Kistiakowsky, W.P. Slichter: Rev. Sci. Instrum. *22*, 333 (1951)
6.17 E.W. Becker, K. Bier: Z. Naturforsch. A *9*, 975 (1954)
6.18 U. Bossel, F.C. Hurlbut, F.S. Sherman: *Proc. 6th Symp. on Rarefied Gas Dynamics*,
      ed. by L. Trilling, H.Y. Wachman (Academic Press, New York 1969)
6.19 G.A. Bird: Phys. Fluids *19*, 1486 (1976)
6.20 Beam Dynamics, 623 E. 57th St., Minneapolis, Minnesota 55417

6.21 H. Ashkenas, F. Sherman: *Proc. 2nd Symp. on Rarefied Gas Dynamics*, ed. by L. Talbot (Academic Press, New York 1960)
6.22 B.B. Hamel, D.R. Willis: Phys. Fluids *5*, 829 (1966)
6.23 E.L. Knuth, S.S. Fisher: J. Chem. Phys. *48*, 1674 (1968)
6.24 D.R. Miller, R.P. Andres: *Proc. 6th Symp. on Rarefied Gas Dynamics*, ed. by L. Trilling, H.Y. Wachman (Academic Press, New York 1969)
6.25 G.A. Bird: Phys. Fluids *11*, 2676 (1970)
6.26 J.P. Toennies, K. Winkelmann: J. Chem. Phys. *66*, 3965 (1977)
6.27 J.B. Anderson, J.B. Fenn: Phys. Fluids *8*, 780 (1965)
6.28 R. Campargue, A. Lebehot, J.C. Lemonnier: *Rarefied Gas Dynamics*, ed. by J.L. Potter (AIAA Publ., New York 1977)
6.29 E.W. Becker, W. Menkes: Z. Phys. *146*, 320 (1956)
6.30 K.R. Way, S.C. Yang, W.C. Swalley: Rev. Sci. Instrum. *47*, 1049 (1976)
6.31 D. Wesner, G. Derry, G. Vidali, T. Thwaites, D.R. Frankl: Surf. Sci. *95*, 337 (1980)
6.32 R. Campargue: Thesis, Paris (1970)
6.33 R. Campargue, A. Lebehot, J.C. Lemonnier, D. Murette: *12th Symp. on Rarefied Gas Dynamics*, Charlottesville, Va., June 1980 (American Institute of Aeronautics and Astronautics, New York, in press)
6.34 W.R. Gentry, L.F. Giese: Rev. Sci. Instrum. *49*, 595 (1978)
6.35 B. Brutschy , H. Haberland: J. Phys. E *13*, 150 (1980)
6.36 J.G. Skoponick, W.M. Pope: Rev. Sci. Instrum. *44*, 76 (1973)
6.37 H. Koschmieder, V. Raible: Rev. Sci. Instrum. *46*, 536 (1975)
6.38 J.W. Hepburn, D. Klimek, K. Lin, J.C. Polyani, S.C. Wallace: J. Chem. Phys. *69*, 431 (1978)
6.39 P.A. Gorry, R. Grice: J. Phys. E *12*, 857 (1979)
6.40 R. Weiss: Rev. Sci. Instrum. *32*, 397 (1961)
6.41 Y.T. Lee, J.D. McDonald, P.R. Le Breton, D.R. Hershbach: Rev. Sci. Instrum. *40*, 1402 (1969)
6.42 R.W. Bickes Jr., R.B. Bernstein: Rev. Sci. Instrum. *41*, 759 (1970)
6.43 B.F. Mason, B.R. Williams: Rev. Sci. Instrum. *43*, 375 (1972)
6.44 D. Hayward: private communication
6.45 A.A. Vostikov, Yu.S. Kusner, A.K. Rebrov, B.E. Semyachkin: Prib. Tekh. Eksp. *1*, 150 (1978)
6.46 M.J. Cardillo, G.I. Becker: Phys. Rev. Lett. *40*, 1148 (1978)
6.47 G. Gallinaro, G. Roba, R. Tatarek: J. Phys. E *11*, 628 (1978)
6.48 H. Nahr, H. Hoinkes, H. Wilsch: J. Chem. Phys. *54*, 3022 (1971)
6.49 M. Faubel, W.M. Holber, J.P. Toennies: Rev. Sci. Instrum. *49*, 449 (1978)
6.50 J.W. McWane, D.E. Oates: Rev. Sci. Instrum. *45*, 1145 (1974)
6.51 The authors are indebted to W. Stocker for this suggestion
6.52 D.J. Auerbach, C.A. Becker, J.P. Cowin, L. Wharton: Rev. Sci. Instrum. *49*, 1518 (1978)
6.53 J. Larscheid, J. Kirschner: Rev. Sci. Instrum. *49*, 1486 (1978)
6.54 J.A. Van den Berg, D.G. Armour: Nucl. Instrum. Methods *153*, 99 (1978)
6.55 J.A. Schwarz, R.J. Madix: Surf. Sci. *46*, 317 (1974)
6.56 G. Frodsham: Rev. Sci. Instrum. *46*, 312 (1975)
6.57 L.A. West, E.I. Kozak, G.A. Somorjai: J. Vac. Sci. Technol. *8*, 430 (1971)
6.58 M.J. Dix, R. Wood, D.H. Slater: J. Phys. E *6*, 1099 (1973)
6.59 T. Engel: Rev. Sci. Instrum. *52*, 301 (1981)
6.60 G. Comsa, R. David, K.D. Rendulic: Phys. Rev. Lett. *38*, 775 (1977)
6.61 B.F. Mason, B.R. Williams: Surf. Sci. *77*, 385 (1978)
6.62 J.M. Horne, D.R. Miller: J. Vac. Sci. Technol. *13*, 351 (1976)
6.63 J.M. Horne, S.C. Yerkes, D.R. Miller: Surf. Sci. *93*, 47 (1980)
6.64 J.J. Cardillo, C.S.Y. Ching, E.F. Greene, G.E. Becker: J. Vac. Sci. Technol. *15*, 423 (1978)
6.65 R.L. Park, J.E. Houston, D.G. Schreiner: Rev. Sci. Instrum. *42*, 60 (1971)
6.66 J.B. Pendry: *Low Energy Electron Diffraction* (Academic Press, London 1974)

6.67 G.C. Wang, M.G. Lagally: Surf. Sci. *81*, 69 (1979)
6.68 G. Comsa: Surf. Sci. *81*, 57 (1979)
6.69 H. Wilsch, L. Greiner, H. Hoinkes, H. Kaarmann, R. Meckler: *Proc. 4th Int. Conf. on Solid Surfaces and 3rd European Conf. on Surface Science*, Cannes, France, Sept. 1980, Suppl. à la Revue "Le Vide, les Couches Minces" Nr. 201 (1980) p. 822

7.1   G. Boato, P. Cantini, L. Mattera: *Proc. 2nd Int. Conf. on Solid Surfaces*, Kyoto, 1974, Jpn. J. Appl. Phys. Suppl. 2, Part 2, p. 553 (1974)
7.2   C. Kittel: *Introduction to Solid State Physics*, 4th ed. (Wiley, New York 1971)
7.3   H. Bilz, W. Kress: *Phonon Dispersion Relations in Insulators*, Springer Series in Solid-State Sciences, Vol. 10 (Springer, Berlin, Heidelberg, New York 1979)
7.4   G.C. Benson, T.A. Claxton: J. Chem. Phys. *48*, 1356 (1968)
7.5   G.E. Laramore, A.C. Switendick: Phys. Rev. B *7*, 3615 (1973)
7.6   G. Benedek, G. Seriani: *Proc. 2nd Int. Conf. on Solid Surfaces*, Kyoto, 1974, Jpn. J. Appl. Phys., Suppl. 2, Part 2, p. 545 (1974)
7.7   G. Benedek: Phys. Rev. Lett. *35*, 234 (1975)
7.8   P. Cantini, G.P. Felcher, R. Tatarek: Phys. Rev. Lett. *37*, 606 (1976)
7.9   P. Cantini, R. Tatarek, G.P. Felcher: Surf. Sci. *63*, 104 (1977)
7.10 N. Garcia, J. Garcia-Sanz, J. Solana: J. Chem. Phys. *66*, 4694 (1977)
7.11 G. Caracciolo, S. Ianotta, G. Scoles, U. Valbusa: J. Chem. Phys. *72*, 4491 (1980)
7.12 H. Frank, H. Hoinkes, H. Wilsch: Surf. Sci. *64*, 362 (1977)
7.13 C.G. Kinniburgh, J.A. Walker: Surf. Sci. *63*, 274 (1977)
7.14 J.A. Walker, C.G. Kinniburgh, J.A.D. Matthew: Surf. Sci. *78*, 221 (1977)
7.15 F.P. Netzer, M. Prutton: J. Phys. C *8*, 2401 (1975)
7.16 A.J. Martin, H. Bilz: *Proc. Int. Conf. on Lattice Dynamics*, Paris, 1977, ed. by M. Balkanski (Flammarion, Paris 1978) p. 327
7.17 J.R. Bledsoe, S.S. Fisher: Surf. Sci. *46*, 129 (1976)
7.18 N. Garcia, G. Armand, J. Lapujoulade: Surf. Sci. *68*, 399 (1977)
7.19 R.G. Rowe, G. Ehrlich: J. Chem. Phys. *63*, 4648 (1975)
7.20 R.G. Rowe, L. Rathburn, G. Ehrlich: Phys. Rev. Lett. *35*, 1104 (1975)
7.21 K.O. Legg, M. Prutton, C.G. Kinniburgh: J. Phys. C *7*, 4236 (1979)
7.22 C.G. Kinniburgh: J. Phys. C *8*, 1382 (1975); *9*, 2695 (1976)
7.23 M. Prutton, J.A. Walker, M.R. Walton-Cook, R.C. Felton, J.A. Ramsey: Surf. Sci. *88*, 95 (1979)

8.1   J.A. Appelbaum, D.R. Hamann: Surf. Sci. *74*, 21 (1978)
8.2   D.J. Chadi: Phys. Rev. Lett. *43*, 43 (1979)
8.3   C.B. Duke, R.J. Meyer, A. Paton, P. Mark, A. Kahn, E. So, J.L. Yeh: J. Vac. Sci. Technol. *16*, 1252 (1979)
8.4   B.J. Mrstik, S.Y. Tong, M.A. Van Hove: J. Vac. Sci. Technol. *16*, 1258 (1979)
8.5   D.J. Chadi: Surf. Sci. *99*, 1 (1980)
8.6   M.J. Cardillo, G.E. Becker: Phys. Rev. B *21*, 1497 (1980)
8.7   M.J. Cardillo, G.E. Becker: Phys. Rev. Lett. *42*, 508 (1979)
8.8   M.J. Cardillo, G.E. Becker, S.J. Sibener, D.R. Miller: Surf. Sci. *107*, 469 (1981)
8.9   W.A. Steele: Surf. Sci. *97*, 478 (1980)
8.10 P. Cantini, G. Boato, R. Colella: Physica B *99*, 59 (1980)
8.11 G. Boato, P. Cantini, R. Colella: Phys. Rev. Lett. *42*, 1635 (1979)
8.12 M.J. Cardillo, G.E. Becker: Surf. Sci. *99*, 269 (1980)
8.13 P.S. Bush, L.M. Raff: J. Chem. Phys. *70*, 5026 (1979)

9.1   N.E. Lang: in *Solid State Physics*, Vol. 28, ed. by F. Seitz, D. Turnbull, H. Ehrenreich (Academic Press, New York 1973) p. 225
9.2   M. Salmeron, R.J. Gale, G.A. Somorjai: J. Chem. Phys. *70*, 2807 (1979)
9.3   G. Boato, P. Cantini, R. Tatarek: *Proc. 7th Int. Vacuum Congr. and 3rd Int. Conf. Solid Surfaces*, Vienna, 1977, ed. by R. Dobrozemsky et al. (American Institute of Aeronautics and Astronautics, New York 1977) p. 1377
9.4   N.R. Hill, M. Haller: to be published
9.5   J. Lapujoulade, Y. Lejay: J. Chem. Phys. *63*, 1389 (1975)

9.6   R. Sau, R.P. Merrill: Surf. Sci. *34*, 268 (1973)
9.7   D.V. Tendulkar, R.E. Stickney: Surf. Sci. *27*, 516 (1971)
9.8   K.H. Rieder, T. Engel: Phys. Rev. Lett. *43*, 373 (1979)
9.9   F.O. Goodman: Surf. Sci. *70*, 578 (1978)
9.10 A.G. Stoll, Jr., J.J. Ehrhardt, R.P. Merrill: J. Chem. Phys. *64*, 34 (1976)
9.11 D.G. Fedak, N.A. Gjostein: Acta Metall. *15*, 827 (1967)
9.12 C.M. Chan, M.A. Van Hove, W.H. Weinberg, E.D. Williams: Solid State Commun. *30*, 47 (1979)
9.13 D.W. Blakely, G.A. Somorjai: Surf. Sci. *65*, 419 (1977)
9.14 S.T. Ceyer, R.J. Gale, S.L. Bernasek, G.A. Somorjai: J. Chem. Phys. *64*, 1934 (1976)
9.15 J. Lapujoulade, Y. Lejay: Surf. Sci. *69*, 354 (1977)
9.16 G. Comsa, G. Mechtesheimer, B. Poelsema, S. Tomoda: Surf. Sci. *89*, 123 (1979)

10.1 B.F. Mason, B.E. Williams: Surf. Sci. *45*, 141 (1974)
10.2 D.O. Hayward, M.R. Walters: Jpn. J. Appl. Phys., Suppl. 2, Pt. 2, p. 587 (1974)
10.3 K.H. Rieder, T. Engel: Phys. Rev. Lett. *45*, 824 (1980)
10.4 T. Engel, K.H. Rieder: Surf. Sci. *109* (in press)
10.5 T. Engel, K.H. Rieder: in preparation
10.6 J. Lapujoulade, Y. Le Cruer, M. Lefort, Y. Lejay, E. Maurel: *Proc. 4th Int. Conf. on Solid Surfaces and 3rd European Conf. on Surface Science*, Cannes, France, Sept. 1980, Suppl. à la Revue "Le Vide, les Couches Minces" Nr. 201 (1980)
10.7 K. Christmann, R.J. Behm, G. Ertl, M.A. Van Hove, W.H. Weinberg: J. Chem. Phys. *70*, 4168 (1979)
10.8 K. Christmann, O. Schober, G. Ertl, M. Neumann: J. Chem. Phys. *60*, 4528 (1974)
10.9 J. Demuth: J. Colloid Interface Sci. *58*, 184 (1977)
10.10 T.N. Taylor, P.J. Estrup: J. Vac. Sci. Technol. *11*, 244 (1974)
10.11 T.H. Upton, W.A. Goddard III: Phys. Rev. Lett. *42*, 472 (1979)
10.12 P.H. Holloway, J.B. Hudson: Surf. Sci. *43*, 141 (1974)
10.13 D.F. Mitchell, P.B. Sewell, M. Cohen: Surf. Sci. *69*, 310 (1977)
10.14 K.H. Rieder: Appl. Surf. Sci. *2*, 74 (1978)
10.15 J.A. Van den Berg, L.K. Verheij, D.G. Armour: Surf. Sci. *91*, 218 (1980)
10.16 L.A. Germer, A.V. MacRae: J. Appl. Phys. *33*, 2923 (1962)

Inelastic Particle-Surface Collisions

Proceedings of the Third International
Workshop on Inelastic Ion-Surface Collisions
Feldkirchen-Westerham, Federal Republic
of Germany, September 17–19, 1980
Editors: E. Taglauer, W. Heiland
1981. 194 figures. VIII, 329 pages
(Springer Series in Chemical Physics,
Volume 17)
ISBN 3-540-10898-X

Contents: Electron Emission. – Electron and
Photon Impact. – Electron Transfer. – Polar-
ized Light Emission. – Excited Particle
Emission. – Index of Contributors.

Secondary Ion Mass Spectrometry SIMS-II

Proceedings of the Second International
Conference on Secondary Ion Mass Spectro-
metry (SIMS II) Stanford University,
Stanford, California, USA,
August 27–31, 1979
Editors: A. Benninghoven, C. A. Evans, Jr.,
R. A. Powell, R. Shimizu, H. A. Storms
1979. 234 figures, 21 tables. XIII, 298 pages
(Springer Series in Chemical Physics,
Volume 9)
ISBN 3-540-09843-7

Contents: Fundamentals. – Quantitation. –
Semiconductors. – Static SIMS. –
Metallurgy. – Instrumentation. – Geology. –
Panel Discussion. – Biology. – Combined
Techniques. – Postdeadline Papers.

Springer-Verlag
Berlin
Heidelberg
New York

M. A. Van Hove, S. Y. Tong

Surface Crystallography by LEED

Theory, Computation and Structural Results

1979. 19 figures, 2 tables. IX, 286 pages
(Springer Series in Chemical Physics,
Volume 2)
ISBN 3-540-09194-7

Contents: Introduction. – The Physics of
LEED. – Basic Aspects of the Programs. –
Symmetry and Its Use. – Calculation of
Diffraction Matrices for Single Bravais-Lattice
Layers. – The Combined Space Method for
Composite Layers: by Matrix Inversion. – The
Combined Space Method for Composite
Layers: by Reverse Scattering Perturbation.
Stacking Layers by Layer Doubling. –
Stacking Layers by Renormalized Forward
Scattering (RFS) Perturbation. – Assembling a
Program: The Main Program and the Input. –
Subroutine Listings. – Structural Results of
LEED Crystallography. – Appendices. –
References. – Subject Index.

Vibrational Spectroscopy of Adsorbates

Editor: R. F. Willis
With contributions by numerous experts
1980. 97 figures, 8 tables. XII, 184 pages
(Springer Series in Chemical Physics,
Volume 15)
ISBN 3-540-10429-1

Contents: Introduction. – Theory of Dipole
Electron Scattering from Adsorbates. – Angle
and Energy Dependent Electron Impact Vibra-
tional Excitation of Adsorbates. – Adsorbate-
Induced Optical Phonons. – Inelastic Elec-
tron Tunnelling Spectroscopy. – Inelastic
Molecular Beam Scattering from Surfaces. –
Neutron Scattering Studies. – Reflection
Absorption Infrared Spectroscopy: Appli-
cation to Carbon Monoxide on Copper. –
Raman Spectroscopy of Adsorbates at Metal
Surfaces. – Vibrations of Monatomic and
Diatomic Ligands in Metal Clusters and
Complexes. – Analogies with Vibrations
of Adsorbed Species on Metals. – Coupling
Induced Vibrational Frequency Shifts and
Island Size Determination: CO on Pt {001}
and Pt {111}.

Electron Spectroscopy for Surface Analysis

Editor: H. Ibach
1977. 123 figures, 5 tables. XI, 255 pages
(Topics in Current Physics, Volume 4)
ISBN 3-540-08078-3

Contents:
H. Ibach: Introduction. – *D. Roy, J. D. Carette:* Design of Electron Spectrometers for Surface Analysis. – *J. Kirschner:* Electron-Excited Core Level Spectroscopies. – *M. Henzler:* Electron Diffraction and Surface Defect Structure. – *B. Feuerbacher, B. Fitton:* Photoemission Spectroscopy. – *H. Froitzheim:* Electron Energy Loss Spectroscopy.

Theory of Chemisorption

Editor: J. R. Smith
1980. 116 figures, 8 tables. XI, 240 pages
(Topics in Current Physics, Volume 19)
ISBN 3-540-09891-7

Contents:
J. R. Smith: Introduction. – *S. C. Ying:* Density Functional Theory of Chemisorption of Simple Metals. – *J. A. Appelbaum, D. R. Hamann:* Chemisorption on Semiconductor Surfaces. – *F. J. Arlinghaus, J. G. Gay, J. R. Smith:* Chemisorption on d-Band Metals. – *B. Kunz:* Cluster Chemisorption. – *T. Wolfram, S. Ellialtioğlu:* Concepts of Surface States and Chemisorption on d-Band Perovskites. – *T. L. Einstein, J. A. Hertz, J. R. Schrieffer:* Theoretical Issues in Chemisorption.

Interactions on Metal Surfaces

Editor: R. Gomer
1975. 112 figures. XI, 310 pages
(Topics in Applied Physics, Volume 4)
ISBN 3-540-07094-X

Contents:
J. R. Smith: Theory of Electronic Properties of Surfaces. – *S. K. Lyo, R. Gomer:* Theory of Chemisorption. – *L. D. Schmidt:* Chemisorption: Aspects of the Experimental Situation. – *D. Menzel:* Desorption Phenomena. – *E. W. Plummer:* Photoemission and Field Emission Spectroscopy. – *E. Bauer:* Low Energy Electron Diffraction (LEED) and Auger Methods. – *M. Boudart:* Concepts in Heterogeneous Catalysis.

X-Ray Optics

Applications to Solids
Editor: H.-J. Queisser
1977. 133 figures, 17 tables. XII, 228 pages
(Topics in Applied Physics, Volume 22)
ISBN 3-540-08462-2

Contents:
H.-J. Queisser: Introduction: Structure and Structuring of Solids. – *M. Yoshimatsu, S. Kozaki:* High Brilliance X-Ray Sources. – *E. Spiller, R. Feder:* X-Ray Lithography. – *U. Bonse, W. Graeff:* X-Ray and Neutron Interferometry. – *A. Authier:* Section Topography. – *W. Hartmann:* Live Topography.

Springer-Verlag Berlin Heidelberg New York